# Building and Wiring Your Trainer Kit
## Instructions for Building the Standard and Micro Automation Enclosures and Wiring your Trainer

By
**Frank Lamb**

© 2020 Frank Lamb, Automation Consulting, LLC. All rights reserved. Except as permitted under the United States Copyright Act of 1976, no part of this publication may be reproduced or distributed in any form or by any means, or stored in a data base or retrieval system, without the prior written permission of the publisher.

Information contained in this work has been obtained from sources believed to be reliable, however the author shall not be responsible for any errors, omissions or damages arising out of use of this information. This work is published with the understanding that Automation Consulting, LLC and the author are supplying information but are not rendering engineering or other professional services. If such services are required, the assistance of an appropriate professional should be sought.

www.automationllc.com

Building and Wiring Your Trainer Kit

# Contents

**About This Book:** ..................................................................................................................... 5

**Enclosure Kit Introduction** ...................................................................................................... 6

**Components** ............................................................................................................................. 7
    **Sizing the Components** ..................................................................................................... 11
    **Mounting Components** ..................................................................................................... 11
        **Drilling and Tapping** ..................................................................................................... 12

**Assembly** ................................................................................................................................. 13
    1. **Unpack** ........................................................................................................................ 13
    2. **Remove Backing from Acrylic** ................................................................................. 13
    3. **End Fasteners** ........................................................................................................... 13
    4. **Connect Extrusion Supports** .................................................................................... 14
    5. **Insert Mesh** ............................................................................................................... 15
    6. **Install Hinges** ............................................................................................................ 15
    7. **Magnetic Catch** ......................................................................................................... 16
    8. **Mesh Retainers** ......................................................................................................... 16
    9. **Install Screws and T-Nuts** ........................................................................................ 17
    10. **Attach Top and Bottom** .......................................................................................... 17
    11. **Drill and Tap Mounting Holes** ............................................................................... 18
    12. **Mount Din-Rail or PLC** ......................................................................................... 18
    13. **Attach Backplane to Supports** ................................................................................ 19
    14. **Attach Handle** ......................................................................................................... 19
    15. **Base Pads** ................................................................................................................. 20
    16. **Attach Device Plate** ................................................................................................. 20
    17. **Install Door** ............................................................................................................. 21
    18. **Door Knob** ............................................................................................................... 21
    19. **Align Door Catch** .................................................................................................... 22
    20. **Tie-Wrap Mesh** ....................................................................................................... 22

**Wiring Kit Introduction** ........................................................................................................ 23

**Components** ........................................................................................................................... 24
    **Wire Lists** ........................................................................................................................... 25
        **Device Cable with Spiral Wrap (A027)** ..................................................................... 25
        **Loose Control Wiring (A028)** ..................................................................................... 26

## Loose Power Wiring (A029) ....................................................................................... 26
# Tools ..................................................................................................................................... 27
# Techniques ........................................................................................................................... 28
## Stripping ............................................................................................................................. 29
## Crimping ............................................................................................................................. 30
# Instructions ......................................................................................................................... 31
## Component Mounting ....................................................................................................... 31
1. **Check Components** ............................................................................................. 31
2. **Check PLC Sizing** ................................................................................................. 31
3. **Base Pads** ............................................................................................................ 32
4. **Handle and Knob** ................................................................................................ 32
5. **Power Button and E-Stop** ................................................................................... 32
6. **Buzzer** .................................................................................................................. 33
7. **Round Buttons** .................................................................................................... 34
8. **Square Indicators** ............................................................................................... 34
9. **Switches** .............................................................................................................. 35
10. **Potentiometers** ................................................................................................... 36
11. **Completed Faceplate Device Mounting** ............................................................ 37
12. **End Anchor and Power TBs** ............................................................................... 38
13. **End Cap** ............................................................................................................... 38
14. **Circuit Breaker** .................................................................................................... 38
15. **PLC** ....................................................................................................................... 39
16. **Power Supply** ...................................................................................................... 40
17. **Fuse Block** ........................................................................................................... 40
18. **DC Terminal Blocks** ............................................................................................ 40
19. **Relay and Socket** ................................................................................................ 40
20. **Wire Duct** ............................................................................................................ 41
## Wiring .................................................................................................................................. 41
## PLC I/O List and Wiring Notes ............................................................................................ 41
21. **AC Wiring** ............................................................................................................. 42
22. **DC Wiring (a)** ....................................................................................................... 43
23. **DC Wiring (b), Output Power** ............................................................................. 45
24. **DC Wiring (c), Input/Output/Analog Commons** ............................................... 45

25. **Potentiometer Butt Splices** .................................................................................. 47
26. **DC Wiring, Door -DC connections** ..................................................................... 47
27. **DC Wiring, Door +DC connections** ..................................................................... 49
28. **DC Extension to Terminal Blocks** ....................................................................... 50
29. **Door to MCR Wiring** ............................................................................................ 51
30. **Door Power/E-Stop I/O Wiring** ........................................................................... 53
31. **Stack Light Indicators I/O Wiring** ....................................................................... 53
32. **Buzzer I/O Wiring** ................................................................................................ 54
33. **Switch I/O Wiring** ................................................................................................ 54
34. **Potentiometer I/O Wiring** ................................................................................... 54
35. **Pushbutton Input Wiring** .................................................................................... 54
36. **Indicator Output Wiring** ..................................................................................... 55
37. **Door Tie-Wrapping/Harness Dressing** ............................................................... 56
38. **I/O Terminations** ................................................................................................. 59
39. **Final Wire Dressing** ............................................................................................. 60
40. **Power Cable** ......................................................................................................... 61

**Check and Apply Power** ................................................................................................. 61

**About the Author** ............................................................................................................ 64

## About This Book:

This book contains two independent sections. The original PLC trainer kit I designed used a completely prefabricated plastic box that I purchased from a local company. The pictures in the wiring section (the second part of this book), are of the original enclosure.

A potential customer ordered a full enclosure and kit in September of 2019, and I shipped it from Tennessee to Massachusetts, after packing it in what I thought was a careful way.

When the customer received the box, the hinge had broken as well as the door latch, so it was back to the drawing board.

Fortunately, I was able to visit the customer and see the damaged enclosure for myself. After discussing it I came up with the kit that is in this book. The advantages of this enclosure kit:

1. It ships flat in a box and is much less likely to be damaged.
2. It is much more rugged and sturdy, with metal edges and stout hinges.
3. The mesh sides allow multiple cables and wires to be connected without drilling holes.

The disadvantages: it is more expensive than my original box, and you have to build it yourself.

Neither this box or the original one are suitable for wet or outdoor areas. I put the hole in the door so that you can see the PLC or other device's lights or indicators.

The original book only covered the wiring of the trainer kit. This one includes the original book and additional instructions on building the enclosure, in two independent sections. There is additional instruction on the tools that are used in the panel building industry, as well as on various techniques. Enjoy building your enclosure, or the full training kit!

# Building your Enclosure Kit

**Small Enclosure**

**Large Enclosure**

## Enclosure Kit Introduction

Thank you for purchasing the Standard or Micro Enclosure kit. This kit includes all of the items required for building a complete training enclosure for different automation components. This is not a NEMA rated enclosure that will protect internal components from water or other environmental effects, it is meant to be used in a lab or training environment.

To make maximum use of this enclosure kit the following requirements should be followed:

1. Smaller components should be DIN rail mountable and fit into the enclosure with enough room for heat dissipation. The din rail can be removed or remounted easily.
2. There are openings in the door of the enclosure and no locking mechanism. If exposed electrical wiring is placed in the enclosure it can harm users.
3. The sides of the enclosure are made of mesh. Water or objects can enter through the mesh and contact electrical components.
4. As with all mechanical assemblies, the proper use of tools is required.

## Components

The following Items are included in your kit. Please ensure that all items are present before starting to build your enclosure.

| Part # | Quantity | Item |
| --- | --- | --- |
| E001 | 2 | Vertical Support (7" Large or 5" Small, tapped ends) |
| E002 | 4 | Horizontal Support (20" Large or 14.5" Small, access holes) |
| E003 | 2 | Hinge with Fasteners |
| E004 | 1 | Door Catch with screws and T-Nuts |
| E005 | 4 | Mesh Retainer (With Support Fasteners E011) |
| E006 | 2 | Side Mesh (Two different sizes) |
| E007 | 4 | End Fastener w/ 1/4-20 screw |
| E008 | 8 | 5/8" 1/4-20 Button Head Screw |
| E009 | 8 | 1/2" 1/4-20 Button Head Screw (With T-Nuts E010) |
| E010 | 8 | 1/4-20 T-Nut (With Screw E009) |
| E011 | 4 | Mesh Retainer Support Fasteners (With Screws E008) |
| E012 | 4 | Door Support Fasteners (With Hinge E003) |
| E013 | 4 | Door Hinge Fasteners (With Hinge E003) |
| E014 | 1 | Backplane (ABS, 20" or 14.5" H, 16" W, 1/2" Thick) |
| E015 | 1 | Top (6" or 8" depth) |
| E016 | 1 | Bottom (6" or 8" depth) |
| E017 | 1 | Door (Two different sizes) |
| E018 | 1 | Device Plate with 6 8-32 Screws and Nuts |
| E019 | 1 | Carry Handle Assembly with 2 Screws |
| E020 | 1 | Door Knob Assembly with Screw and Fender Washer |
| E021 | 4 | Adhesive Base Pads |
| E022 | 2 | Din Rail with 4 8-32 screws |
| E023 | 1 | Ball End Hex Allen Wrench |

In addition to the Ball End Hex Allen Wrench listed above, you will need various hand tools; an adjustable wrench, screwdriver, 3/32" Allen Wrench and possibly a drill and tap set if you plan to place the din rail in a different position or surface mount a large object such as a rack based PLC.

The PLC Trainer kit includes all of the components to wire a PLC into this enclosure and is marketed as a separate item.

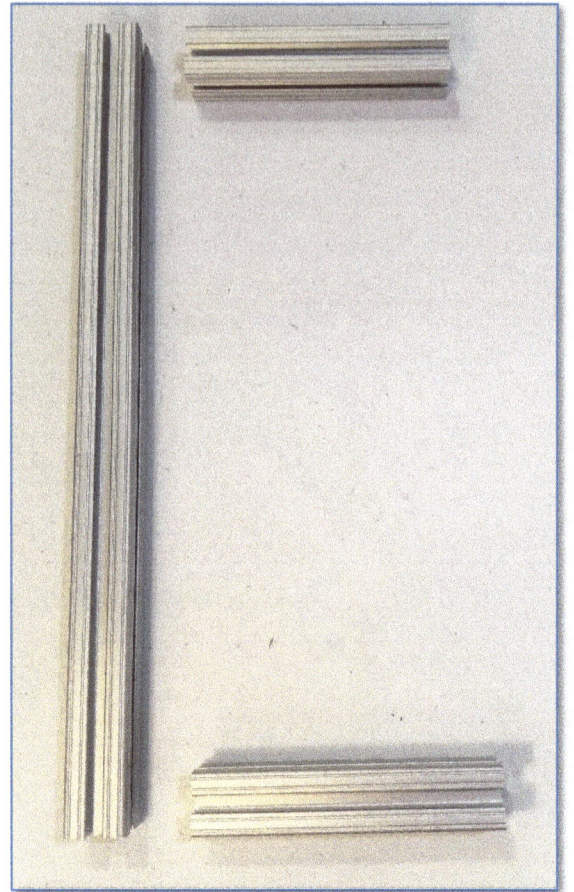

*Figure 1 - E001, E002 Aluminum Extrusion*

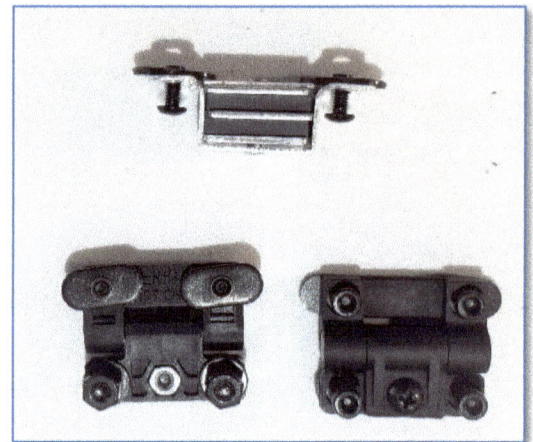

*Figure 2 - E003, E004 Hinge and Door Catch*

*Figure 3 - E005 Mesh Retainers*

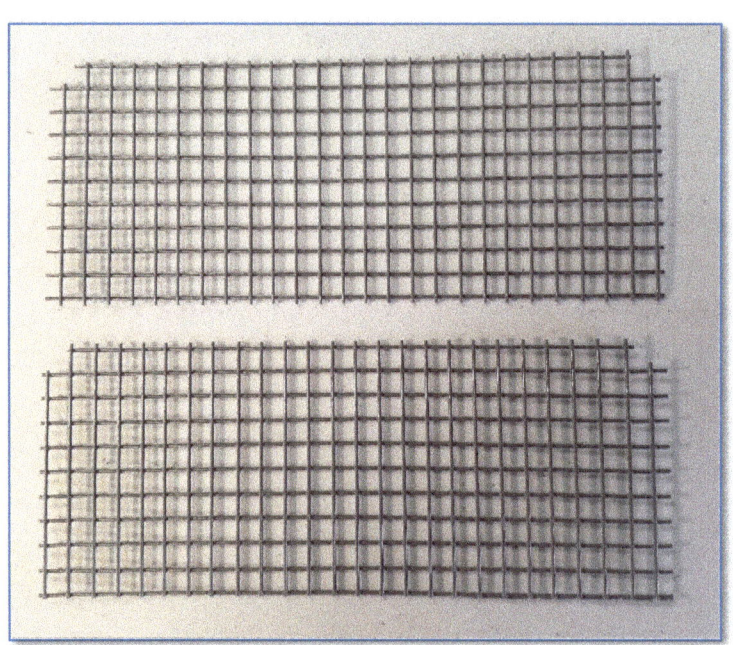

*Figure 4 – E006 Mesh (Sides)*

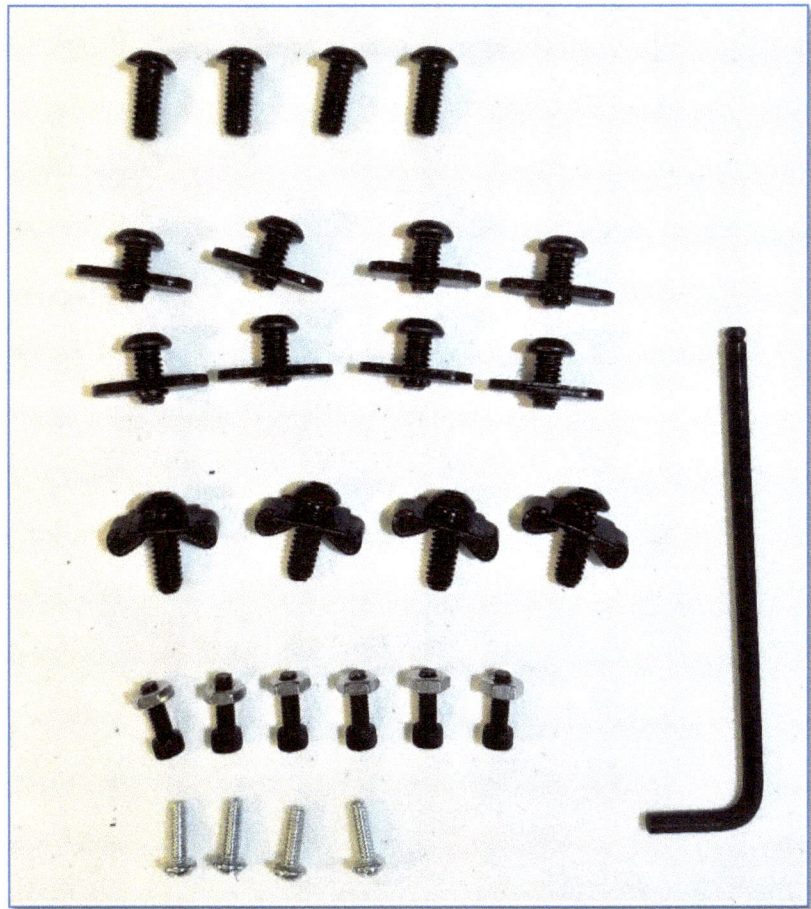

**E008** – 5/8" x ¼-20 Screws for Extrusion to Backplane attachment

**E009** – 1/2" x ¼-20 Screws with T-Nuts (E010) for Top and Bottom attachment to Extrusion

**E007** – End Fasteners for Extrusion to Extrusion attachment

**E018 a1-f1 and a2-f2** – 8-32 screws and nuts for attachment of device plate E018 to door E017

**E022 a-d** – 8-32 screws for attachment of Din Rail E022 to Backplane E014

*Figure 5 - E007, E008, E009, E010, E023 Fasteners and Allen Wrench*

*Figure 6 - E014 Backplane*

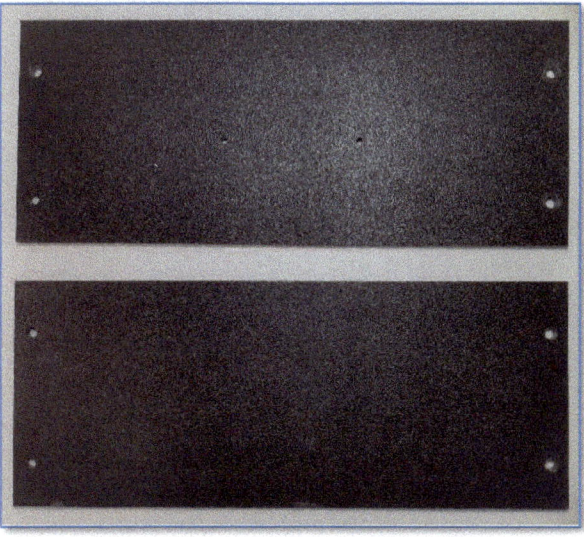

*Figure 7 - E015, E016 Top and Bottom*

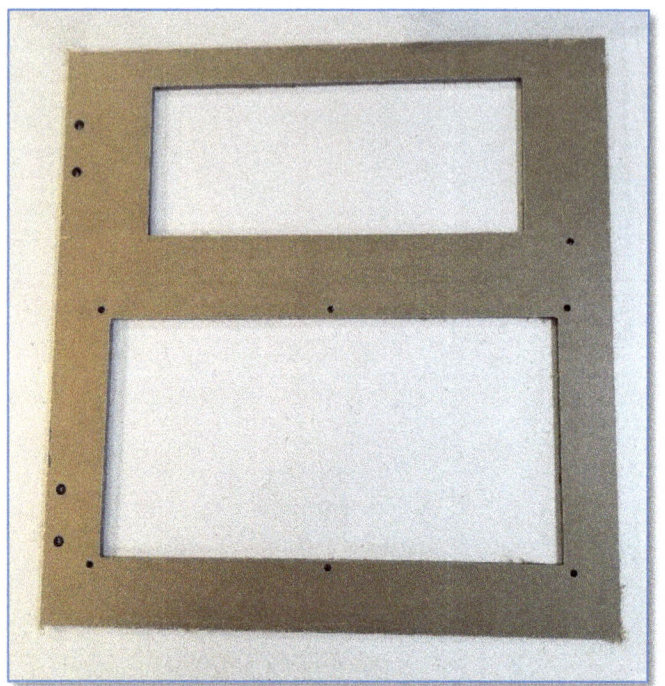

Figure 8 - E017 Door (with adhesive paper cover)

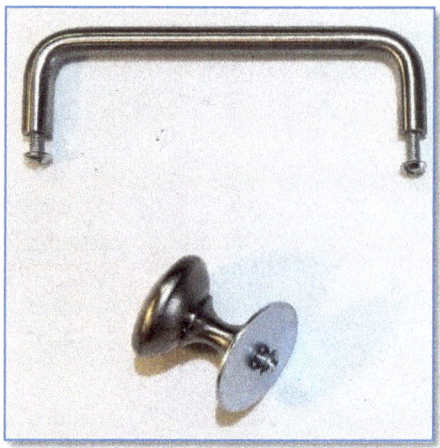

Figure 9 - E019, E020 Handle and Knob

Figure 10 - E021 Adhesive Base Pads

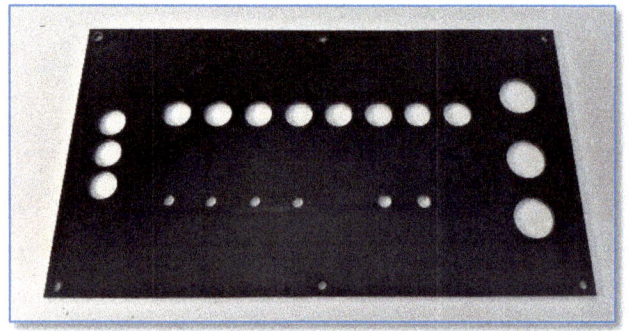

Figure 11 - E018 Device Plate

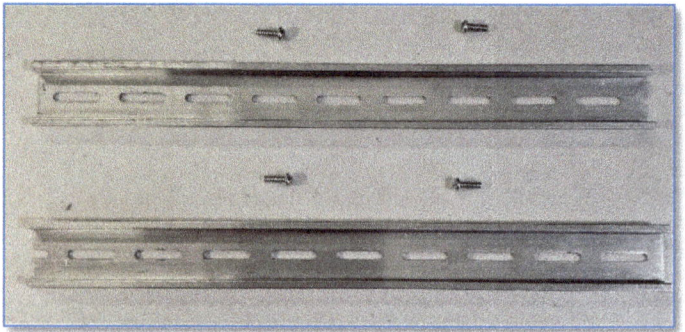

Figure 12 - E022 Din Rail and Screws

## Sizing the Components

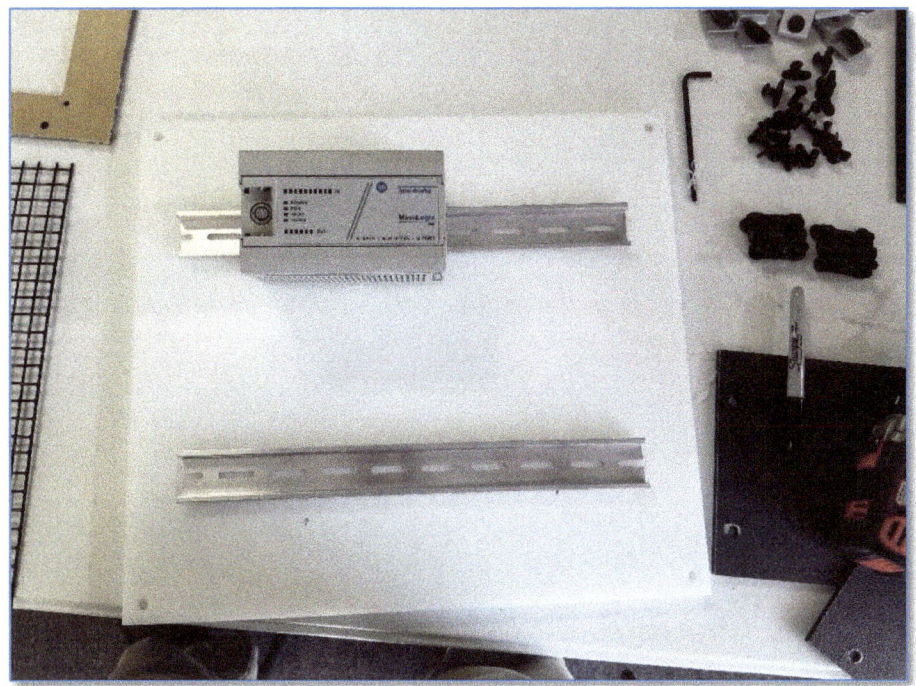

*Figure 13 – Component Location*

Figure 13 above shows a PLC place over a piece of din rail to show the tentative location of a component. The white ABS backplane (E014) is pre-drilled and tapped to mount the two pieces of din rail (E022). It is possible that in the case of the larger kit you may wish to mount a large component such as a PLC rack to the backplane. In this case you will need to mark hole positions and then drill and tap new holes for mounting. After locating the item to be mounted on the backplane, a marker can be used to mark the spot where the holes are to be drilled through the mounting holes themselves.

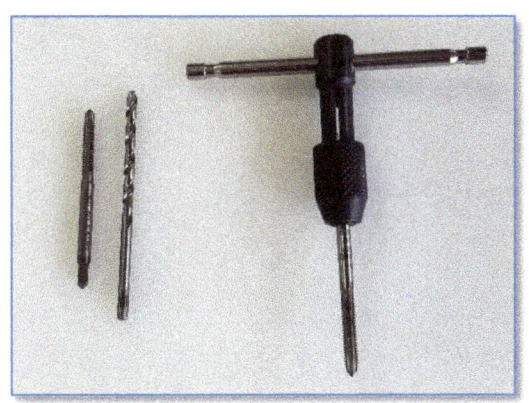

## Mounting Components

After marking the holes and removing the item from the backplane, use a drill and tap set to create the tapped holes for mounting components.

*Figure 14 - Drill and Tap with Handle*

## Drilling and Tapping

After marking where the tapped holes will be located, drill a hole all the way through the backplane. Ensure that there is nothing behind the backplane that can be damaged while drilling.

It is important to ensure that the drill bit is perpendicular to the backplane to ensure a straight hole.

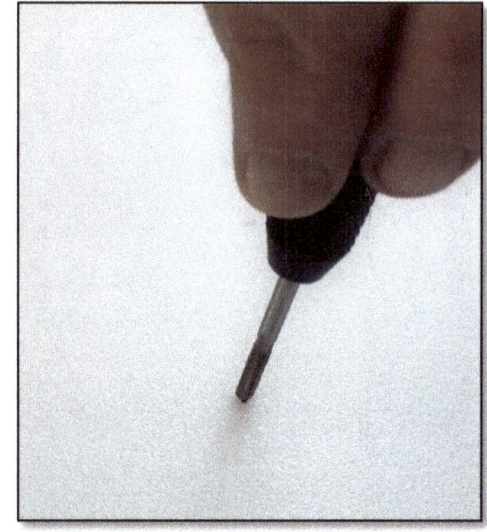

*Figure 15 - Drill Hole w/#29 bit*

*Figure 16 - Hand tap hole w/ 8-32 tap*

Using a tap of the appropriate size, insert the tap into the hole and turn it clockwise, occasionally backing it out to clear debris. Tap the hole all the way through so that the mounting screws can be threaded completely into the hole.

If using din rail, the rail can be left attached to the backplane while the enclosure is assembled.

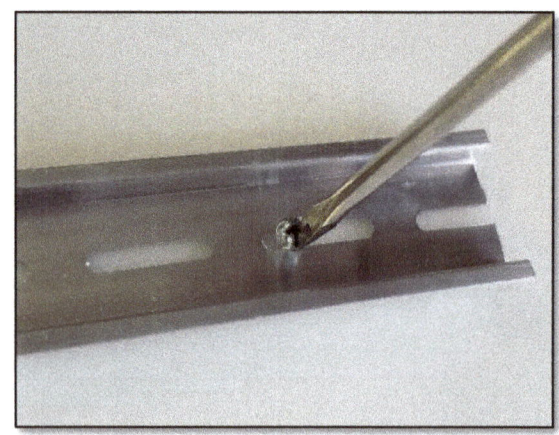

*Figure 17 - Tap Through Backplane*

*Figure 18 - Attach Din Rail*

## Assembly

1. Unpack – Ensure that all items listed in the components list are present. Some items listed separately are attached together.

2. Remove Backing from Acrylic - Carefully peel the adhesive backing from the Door (E017) and the Device Plate (E018) and discard it.

Figure 19 - Remove Adhesive Backing

3. End Fasteners – Screw the End Fasteners (E007) into one end of each of the four Horizontal Support pieces (E002) using the included Ball End Allen Wrench (E023)

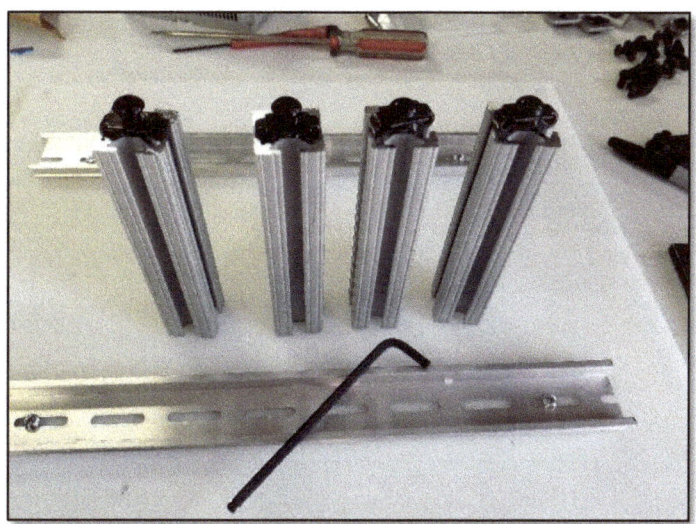

Figure 20 - Install End Fasteners

4. Connect Extrusion Supports – Slide the Horizontal Support (E002) with End Fastener into the slot of the Vertical Support (E001) with the access hole lined up with the screw. Insert the ball End Allen Wrench through the hole and tighten the screw.

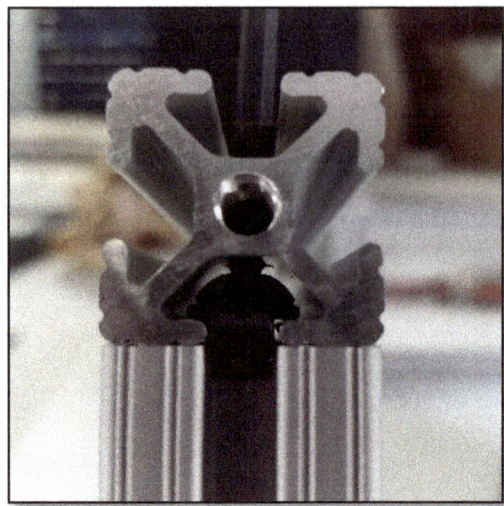
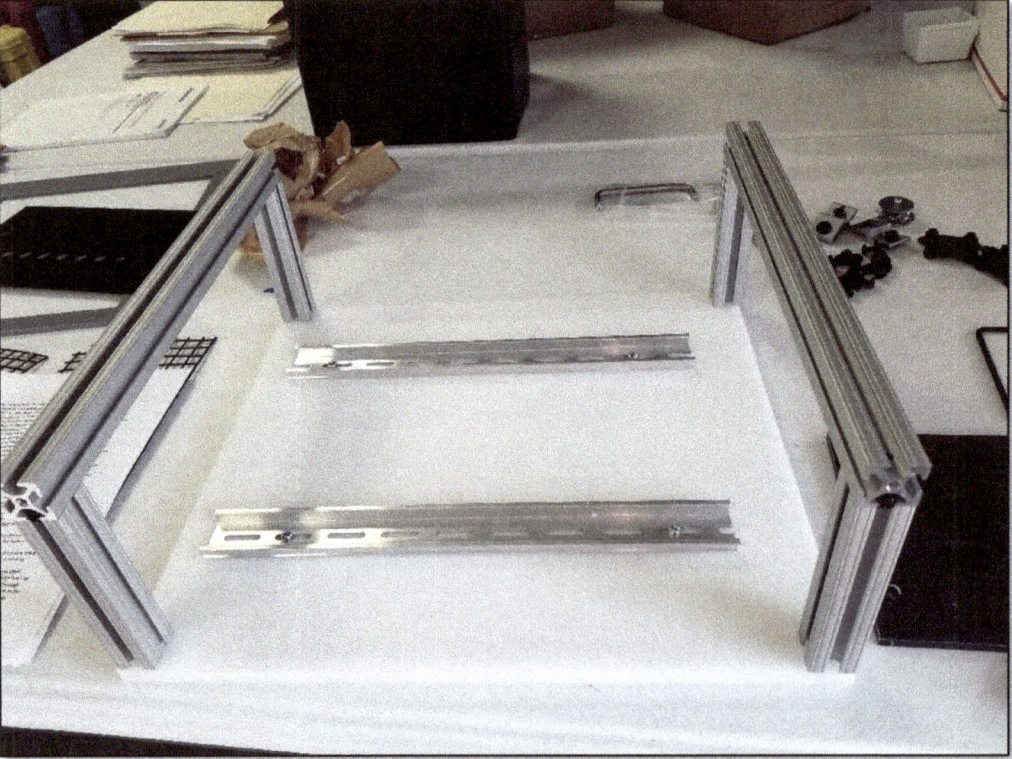

*Figure 21 - Connection of Extrusion Supports*

5. Insert Mesh – Slide the Side Mesh (E006) pieces into the slots of the extrusion support assembly so that the clipped corners of the mesh are in the corners of the extrusion.

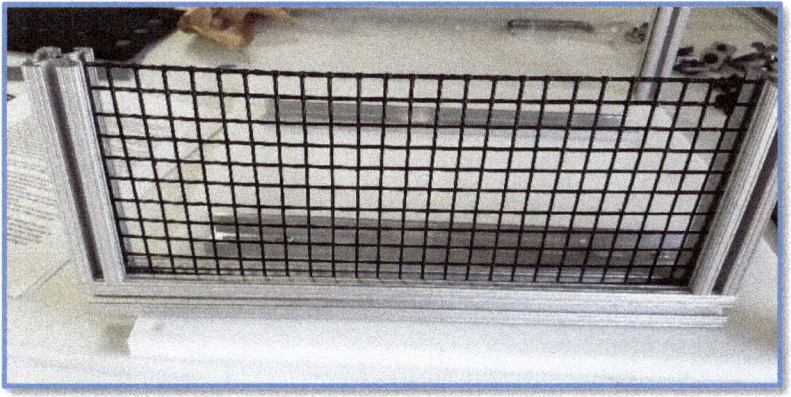

*Figure 23 - Corner*

*Figure 22 - Mesh in Slot*

6. Install Hinges - Slide both of the two Hinges' E003 T-Nuts into the outside slot of the vertical support piece of one of the assemblies. This becomes the left side assembly. Loosely tighten screws to ensure hinges don't slide. Location is not yet important. The other side of the hinge with the bolt and nut now face the inside edge of the assembly.

*Figure 24 - Hinge Insertion*

7. Magnetic Catch - Slide the Magnetic Catch E004 T-Nuts into the inside slot of the right-side extrusion and mesh assembly with the magnet facing toward the outside. Loosely tighten screws to ensure the catch doesn't slide. Location is not yet important.

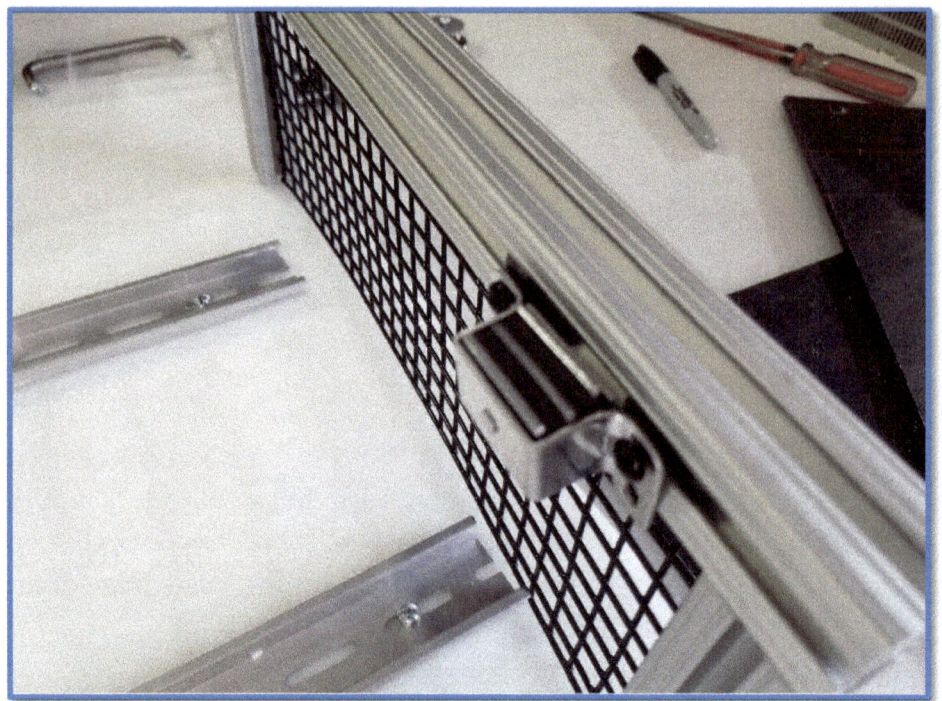

*Figure 25 - Insert Door Catch*

8. Mesh Retainers - Remove the square part of the mesh retainer E005, remove the screw E008 from the slotted hole and set aside. Slide the mesh retainer T-Nuts into the outside top and bottom slots of both the left and right assembles, with the slotted hole over the mesh. Place the screw through the slotted hole and mesh and rethread the square part of the mesh retainer onto the screw on the inside of the mesh. Ensure that the mesh retainers are in even central positions on the mesh.

*Figure 26 - Mesh Retainers*

9. Install Screws and T-Nuts - Place a Screw E009 through each hole on the Top E015 and bottom E016 pieces. Thread a T-Nut E010 onto each screw.

*Figure 27 – Insert Screws and Thread T-Nuts*

10. Attach Top and Bottom - Slide the Top E015 into the top slots of the left and right-side assemblies. The handle holes are closer to the back side of the top than to the front. Slide the top as far toward the front as it will go and tighten the screws. Slide the Bottom E016 into the bottom slots of the left and right-side assemblies. Slide the bottom as far toward the front as it will go and tighten the screws.

*Figure 28 – Slide Top and Bottom into slots*

11. Drill and Tap Mounting Holes – The backplane is made of High-Density Polyethylene (HDPE) and can easily be drilled and tapped. Holes have already been drilled and tapped for the included din-rail, if you wish to use this, skip to the next step. If not, mark the holes on the object you wish to mount using a felt tip pen. Drill holes at the marked points using the appropriate drill size.

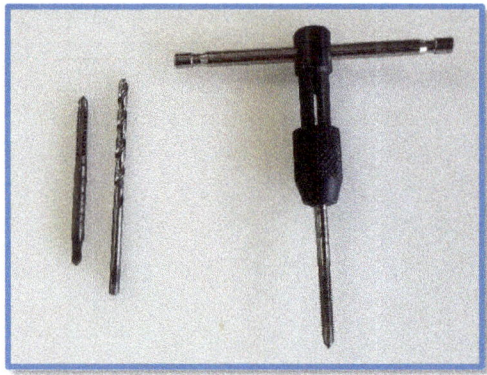

Insert the tap and turn clockwise, periodically unscrewing it to clear the threads. Tap the holes all the way through the backplane.

*Figure 29 – Drill and Tap Mounting Holes*

12. Mount Din-Rail or PLC – Using the 8-32 screws E022 a-d, attach the two pieces of Din Rail E022 to the backplane. If you are mounting a different object such as a PLC to the backplane, you can mount it now or wait until the next step when the backplane is attached to the supports.

*Figure 30 – Mount Din-rail*

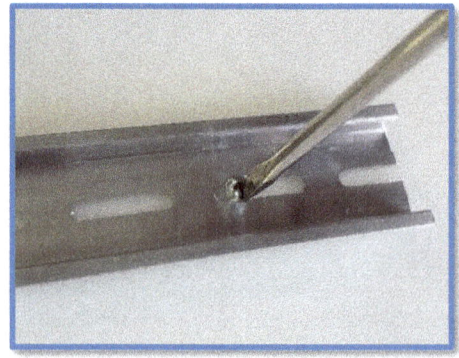

13. Attach Backplane to Supports - Place the four screws E008 through the countersunk holes on the back of the Backplane E014. Screw them into the ends of the horizontal support pieces.

*Figure 31 – Attach Backplane*

*Figure 32 – After Step 13*

14. Attach Handle - Remove the 8-32 screws from the Handle E019 and insert them through the holes in the top from the inside. Thread them into the handle E019 and tighten.

*Figure 33 – Handle*

15. **Base Pads** - Remove the backing and apply the Adhesive Pads E021 to the bottom of the enclosure towards the corners. You can now pick the enclosure up and set it down easily without scratching anything.

*Figure 34 – Base Pads*

16. **Attach Device Plate** - Ensure that the paper backing has been removed from the Device Plate E018 and the Door E017. Place the screws E018 a1-f1 through the holes in the device plate with the 3 large holes to the right and the 6 small holes to the bottom as shown in Figure 36. Place the screws through the holes around the large opening in the door E017, ensure that the four hinge holes are to the left side of the door as shown in Figure 35. Thread the 8-32 nuts onto the screws and tighten.

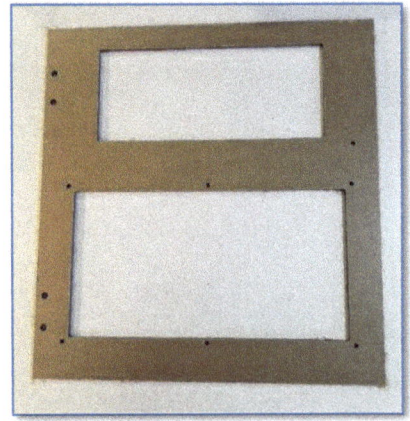

*Figure 35 – Door*

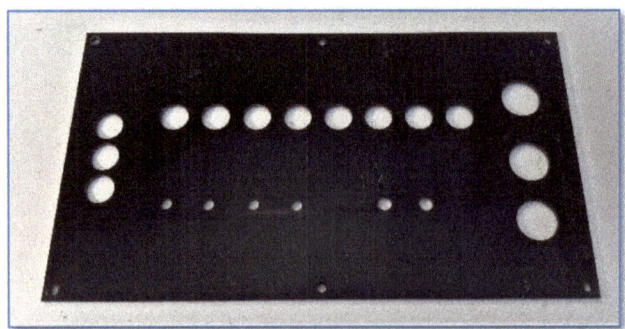

*Figure 36 – Device Plate*

17. Install Door - Loosen the hinges in the slots on the left side of the enclosure. Remove the nut and bolt from the hinges' holes. Align the holes with the holes in the door, with the device plate to the outside and in the lower half of the door. Place the bolts through the hinge holes and through the door holes and thread the nuts onto the bolts on the inside of the door. Tighten the nuts. Slide the hinges in the slot until the door is in the proper position, fitting into the top, bottom, left and right sides.

*Figure 37 – Installed Door*

18. Door Knob - Remove the screw and fender washer from the Door Knob E020. Insert the screw through the washer, insert through the hole on the right side of the door from the back. Thread the knob onto the screw.

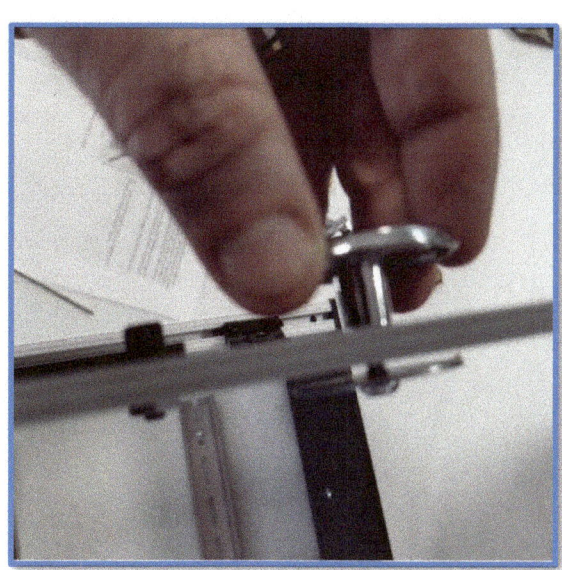

*Figure 38 – Knob and Washer*

19. Align Door Catch - Loosen the Door Catch E004 in the slot and slide it until it aligns with the fender washer and knob E020. Tighten it in place.

*Figure 39 – Completed Assembly with Latch aligned*

20. Tie-Wrap Mesh - Insert Tie-wrap as shown in Figure 40. Tighten and cut off excess. This ensures mesh doesn't move or rattle.

   **Enclosure is complete!**

*Figure 40 – Tie-wrap mesh to backplane*

# Wiring your PLC Trainer Kit

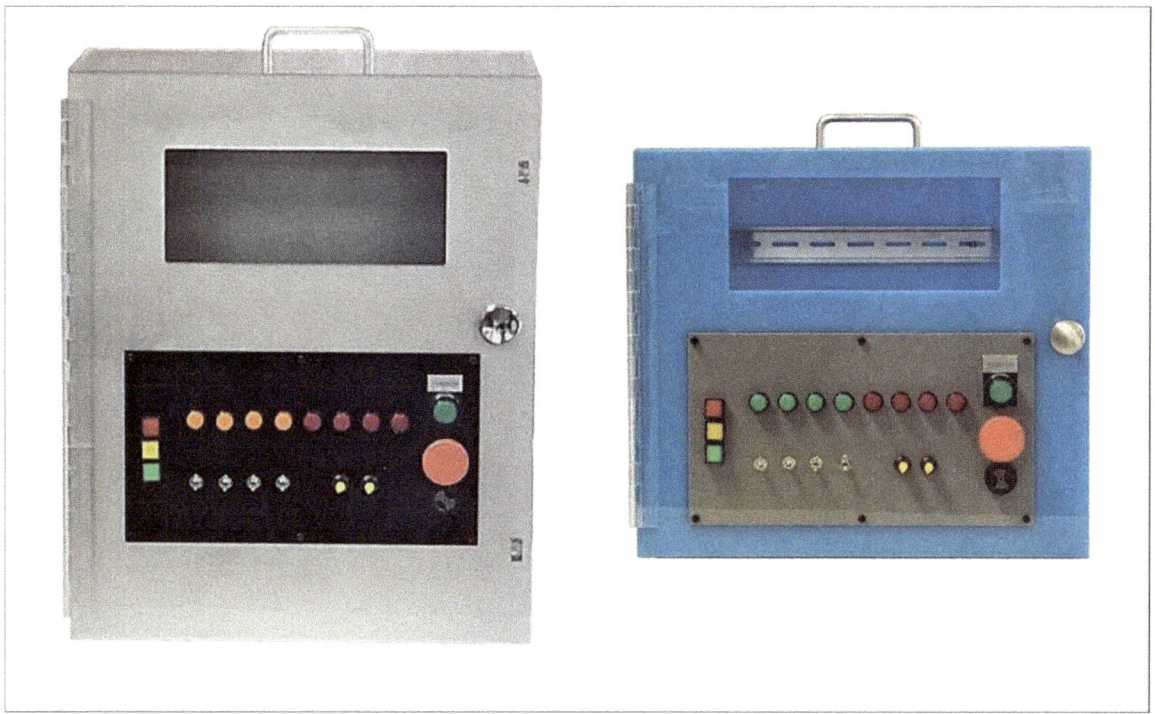

Standard Trainer Kit          Micro Trainer Kit

## Wiring Kit Introduction

Thank you for purchasing the Standard or Micro Trainer kit. This kit includes all of the items required for building a complete PLC trainer with 8 lighted pushbuttons, 4 switches and two potentiometers. In addition, a complete Emergency Stop circuit, a stack light simulation, and a buzzer are included to simulate real industrial applications. It can be used with various PLCs.

To make maximum use of this trainer kit the following requirements should be followed:

5. The PLC should be either 120VAC or 24VDC powered. The power cord fits a US standard 120 V electrical outlet.
6. There should be a minimum of 15 DC inputs available.
7. There should be a minimum of two analog inputs available.
8. There should be a minimum of 12 DC or relay outputs available. There should also be a separate set of commons for the 8 pilot lights so that they can be switched by the Master Control Relay (MCR). The stack lights and buzzer are unswitched.

## Components

The following items are included in your kit, (only one enclosure kit is included). Please ensure that all listed items are present before starting to build your panel.

| Part # | Quantity | Item |
|---|---|---|
| A001S | 1 | Enclosure Kit (Standard) |
| A001M | 1 | Enclosure Kit (Micro) |
| A002 | 1 | Device Plate (Included with Enclosure Kit, different models and colors available) |
| A003 | 1 | Power Supply (Rhino PSL-24-060, 24VDC, 2.5A) |
| A004 | 1 | Circuit Breaker (Eaton FAZ-C4-1-SP-NA, 4A) |
| A005 | 8 | Terminal Blocks (Auto. Dir. KN-T12) |
| A006 | 3 | End Caps (Auto. Dir. KN-ECT6) |
| A007 | 5 | End Anchors (Auto. Dir. KN-EB4) |
| A008 | 1 | Fuse Terminal Block (Auto. Dir. DN-F6L24) |
| A009 | 1 | Cartridge Fuse (Edison AGC3, 3A) |
| A010 | 2 | Terminal Block Center Jumpers (Auto. Dir. KN-3J12) |
| A011 | 1 | Wire Duct (Iboco DN35AG-1) |
| A012 | 1 | 3 Prong Pigtail (Husky 165-018) |
| A013 | 1 | Green Lighted 22mm Pushbutton, 1 NO Contact (Auto. Dir. GCX3202-24) |
| A014 | 1 | Red Twist 22mm E-Stop, 1 NC Contact (Auto. Dir. GCX3131) |
| A015 | 1 | Piezo Buzzer 22mm (Auto. Dir. ECX2070-24) |
| A016 | 2 | NO Pushbutton Contact Blocks (Auto. Dir. ECX1040) |
| A017 | 1 | Power On Legend Plate (Auto Dir. ECX1670A-B21) |
| A018 | 3 | Square 16mm Lighted Maintained Buttons R-Y-G (Baomain 24VDC) |
| A019 | 8 | Round 16mm Lighted Momentary Buttons, various colors (Baomain 24VDC) |
| A020 | 4 | SPDT Micro Mini Switch (Gadgeteer) |
| A021 | 2 | 1K Potentiometer (Uxcell) with B/R/W leads |
| A022 | 1 | DPDT 24V Relay (Idec RY2S-U DC24V) |
| A023 | 1 | DPDT Relay Socket (Idec SY2S-05) |
| A024 | 1 | Enclosure Carry Handle with Screws |
| A025 | 1 | Enclosure Door Knob with Screws |
| A026 | 4 | Enclosure Adhesive Base Pads |
| A027 | 1 | Device Cable (48", 25 Wires, 22AWG) with Spiral Wrap |
| A028 | 1 | Loose Signal Wires (48", 11 Wires, 20AWG) |
| A029 | 1 | Loose Power Wires (Various Lengths, 5 Wires, 16-20AWG) |
| A030 | 12 (14) | Insulated 0.108 F Spade Connectors |
| A031 | 34 (40) | Uninsulated 0.108 F Spade Connectors |
| A032 | 4 (5) | Insulated Butt Splice Connectors |
| A033 | 20 | Tie Wraps |

## Wire Lists

These lists are the wire bundles A027, A028 and A029 included in your kit, their size and purpose. More detailed information is included in the assembly instructions.

### Device Cable with Spiral Wrap (A027)

| Wire # | Device Cable | Length | Color | Device | I/O Ex. (1100) |
|---|---|---|---|---|---|
| D001 | 22 AWG Cable | 48" | Black | PB1 | I 0/0 |
| D002 | 22 AWG Cable | 48" | Brown | PB2 | I 0/1 |
| D003 | 22 AWG Cable | 48" | Red | PB3 | I 0/2 |
| D004 | 22 AWG Cable | 48" | Orange | PB4 | I 0/3 |
| D005 | 22 AWG Cable | 48" | Yellow | PB5 | I 0/4 |
| D006 | 22 AWG Cable | 48" | Green | PB6 | I 0/5 |
| D007 | 22 AWG Cable | 48" | Blue | PB7 | I 0/6 |
| D008 | 22 AWG Cable | 48" | Violet | PB8 | I 0/7 |
| D009 | 22 AWG Cable | 48" | Gray | Ind1 | O 2/0 |
| D010 | 22 AWG Cable | 48" | White | Ind2 | O 2/1 |
| D011 | 22 AWG Cable | 48" | Pink | Ind3 | O 2/2 |
| D012 | 22 AWG Cable | 48" | Tan | Ind4 | O 2/3 |
| D013 | 22 AWG Cable | 48" | Red/Black | Ind5 | O 2/4 |
| D014 | 22 AWG Cable | 48" | Red/Yellow | Ind6 | O 2/5 |
| D015 | 22 AWG Cable | 48" | Red/Green | Ind7 | O 2/6 |
| D016 | 22 AWG Cable | 48" | White/Black | Ind8 | O 2/7 |
| D017 | 22 AWG Cable | 48" | White/Red | SW1 | I 0/8 |
| D018 | 22 AWG Cable | 48" | White/Orange | SW2 | I 0/9 |
| D019 | 22 AWG Cable | 48" | White/Yellow | SW3 | I 1/0 |
| D020 | 22 AWG Cable | 48" | White/Green | SW4 | I 1/1 |
| D021 | 22 AWG Cable | 48" | White/Blue | Pot1 | I V1 |
| D022 | 22 AWG Cable | 48" | White/Black/Red | Pot2 | I V2 |
| D023 | 22 AWG Cable | 48" | White/Brown | E-Stop PB | I 1/2 |
| D024 | 22 AWG Cable | 48" | White/Violet | Reset PB | I 1/3 |
| D025 | 22 AWG Cable | 48" | White/Gray | <Spare> | |

## Loose Control Wiring (A028)

| Wire # | Loose Wire | Length | Color | Device | I/O Ex. (1100) |
|---|---|---|---|---|---|
| L001 | 20 AWG | 48" | Black | Buzzer | O 0/3 |
| L002 | 20 AWG (2) | 48" | Brown | Door + DC | |
| L003 | 20 AWG | 48" | Red | Red Stack | O 0/2 |
| L004 | 20 AWG | 48" | Orange | Pin 5 MCR (SW) | I 1/4 |
| L005 | 20 AWG | 48" | Yellow | Yellow Stack | O 0/1 |
| L006 | 20 AWG | 48" | Green | Green Stack | O 0/0 |
| L007 | 20 AWG | 48" | Blue | Door to Brown | |
| L008 | 20 AWG | 48" | Violet | Pin 8 MCR | |
| L009 | 20 AWG | 48" | Gray | Door -DC | |
| L010 | 20 AWG | 48" | White | Pin 14 MCR | |

## Loose Power Wiring (A029)

| Wire # | Power Wire | Length | Color | Device | I/O Ex. (1100) |
|---|---|---|---|---|---|
| P001 | 16 AWG MTW | 24" | Red | AC Power L1 | PLC/PS |
| P002 | 16 AWG MTW | 24" | White | AC Power N | PLC/PS |
| P003 | 16 AWG MTW | 24" | Green | AC Power Gnd. | PLC/PS |
| P004 | 20 AWG MTW | 48" | Blue | DC + | |
| P005 | 20 AWG MTW | 48" | Blue/White | DC - | |

The instructions for this PLC Trainer Kit were written using an Allen-Bradley MicroLogix 1100 PLC, Part# 1763-L16BWA and two expansion cards, a 1762-IQ8 DC Input Card and a 1762-OB8 DC Output card. This PLC has 10 DC inputs and 6 relay outputs built into the chassis, along with two analog voltage/current inputs. The input card has 8 DC inputs and the output card has 8 DC outputs.

To use this trainer as designed, it is necessary that the VDC terminals for the outputs are separable into groups. There are 4 unswitched outputs (stack light and buzzer) and 8 switched outputs (the pilot lights). This simulates real world industrial systems, where moving items (simulated by the pilot lights), are disabled by the Emergency Stop/MCR circuit.

This example was also made with a Micro Trainer, which is an appropriate size for smaller din-rail mounted PLCs. The larger Standard Trainer is deeper and taller than the Micro Trainer. It is appropriate for larger rack-type PLCs.

## Tools

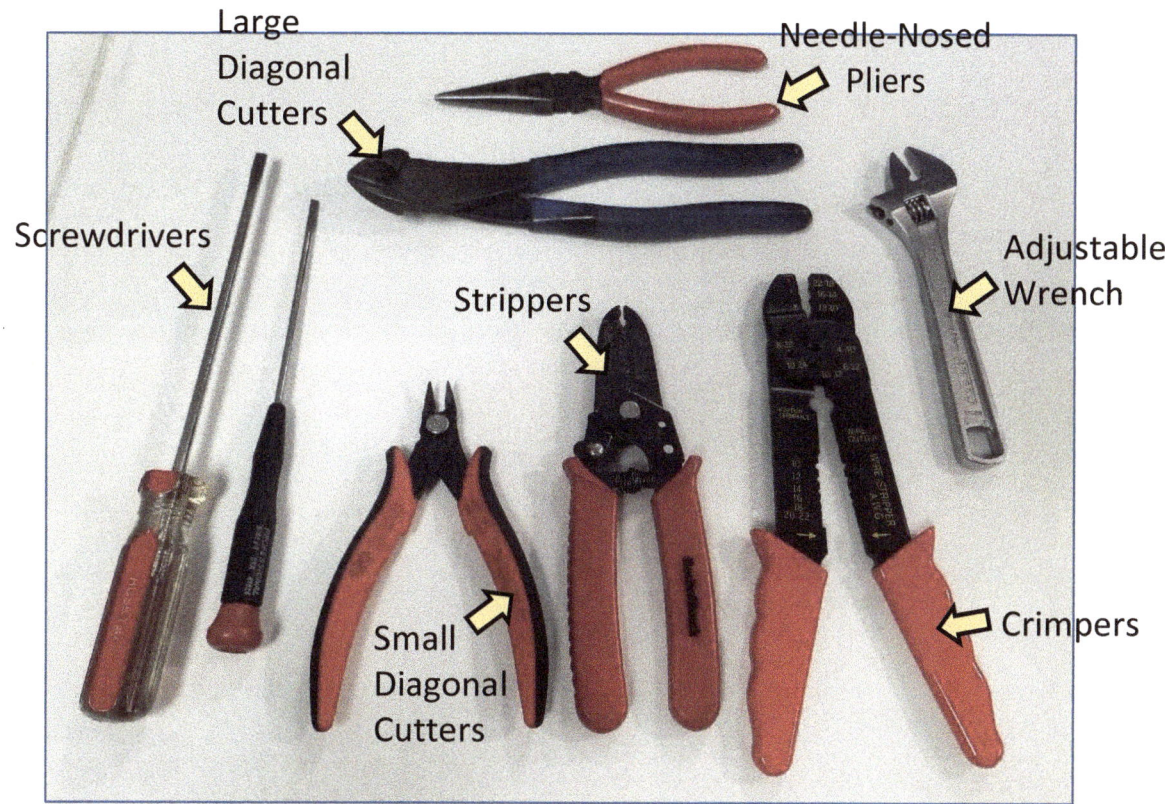

Shown above are some of the basic tools needed to build a control panel. Other optional tools include vise grips, alligator clips, a soldering iron and solder.

There are also higher quality tools available if you can afford them. This kit does not require and does not include ferrules, but placing ferrules on the stripped ends of wire makes them less susceptible to breakage. It can also help to hold twisted wires together as shown below.

The tool used to crimp the ferrule in the pictures above is an expensive ratcheting type, but good results can be obtained by using pliers or vise grips.

Shown below are some of the better, higher-priced tools available. Left to right: Strippers, Ratcheting Crimpers, and Ferrule Crimpers.

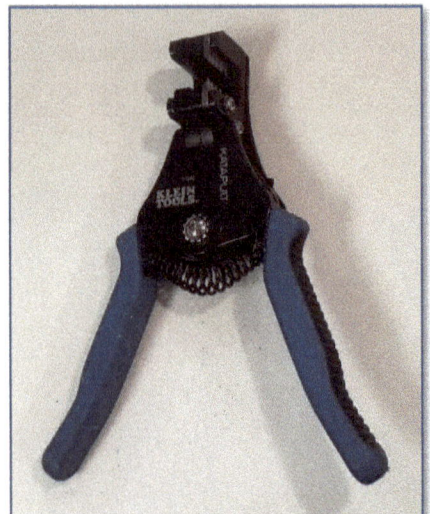 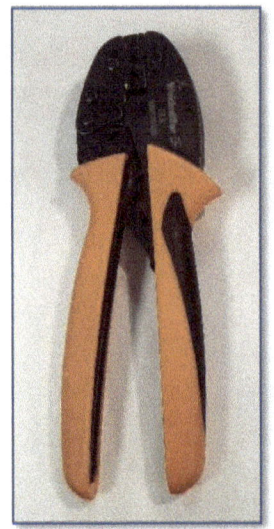 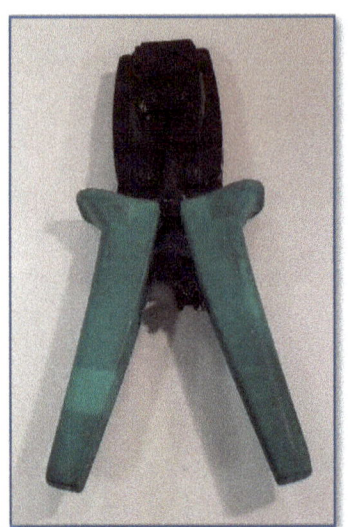

## Techniques

Properly wiring a control panel requires a good eye for distance and right angles. Wires should be stripped to the proper length so that bare wire does not extend outside of the terminal block or ferrule if used.

Usually wires are labeled according to a set of schematics with wire numbers or I/O designations. This kit does not include labels, instead the wires are color coded. When labels are used, it is important for them to be readable from the front and all the same direction. They should also be the same distance from the terminal block or termination point.

 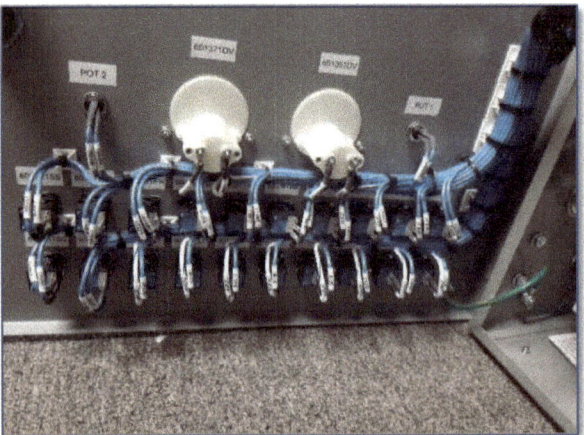

The pictures above show heat shrinkable labels. Notice also that the terminal blocks are labeled with the same numbers.

These kits only have a few terminal blocks and wires are color coded. As mentioned before, ferrules are optional.

## Stripping

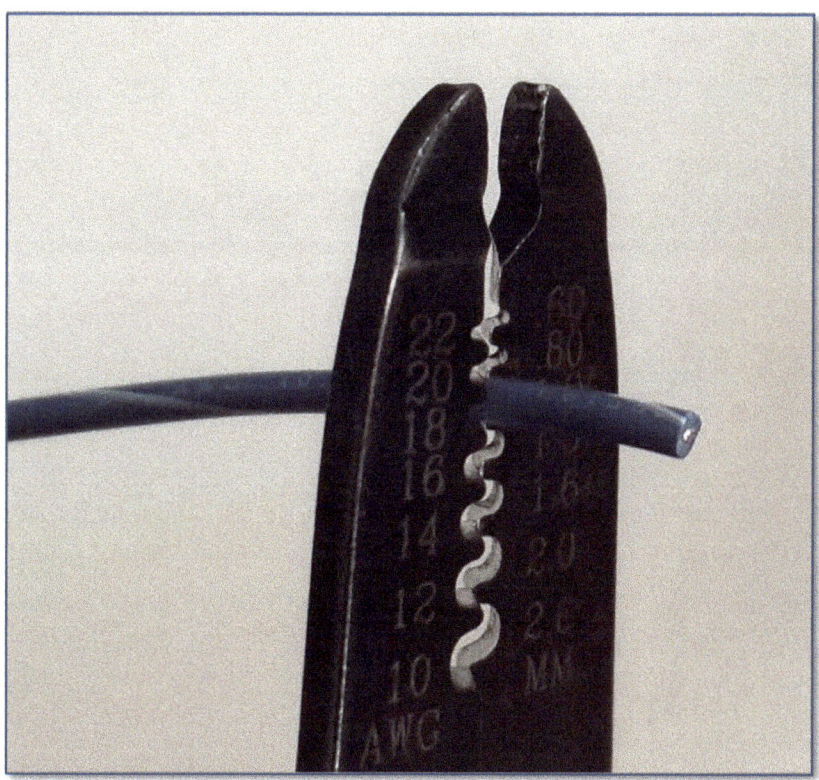

When stripping wire, it is important to use the proper size on the stripping tool. The picture above shows the labels on the strippers corresponding to the gauge of the wire.

The picture to the left shows a wire with ferrule inserted into a terminal block. As you can see, there is no bare wire extending outside of the block.

The picture to the right shows a properly stripped and twisted wire before inserting it under the screw terminal of a relay. Very little of the bare wire will extend outside of the terminal.

## Crimping

Crimpable terminals have been included for attaching wires to some of the devices mounted in the door. It is important that crimps are made firmly so that the wires can't slip out of the terminals.

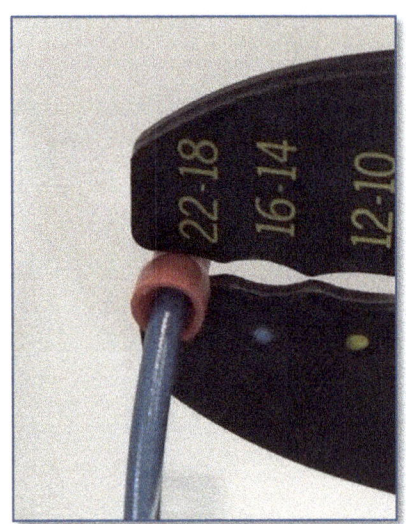

Like strippers, crimpers are sized according to the type of crimp terminal being used. The red spade terminals included in this kit are all used on 20AWG and 22 AWG wire.

Red butt splices are also used to attach wires to the potentiometers.

## Instructions
### Component Mounting

1. **Check Components** - Ensure that all components are present in the kit. Carefully unpack items from inside the enclosure and set them aside.

2. **Check PLC Sizing** - Place your PLC inside of the enclosure and ensure that it fits. If it is a din-rail mount unit, set it on the left side of the din-rail. If it fits properly, do not mount it yet, set it aside. If it not a din-rail mount PLC, remove the din-rail and place your PLC inside the enclosure, near the top. Mark the mounting holes for your PLC. Use a plastic bit to drill and tap holes for your PLC, alternatively you can use bolts with a nut and washer on the back side.

**Note:** Drilling plastic requires a special bit as shown in the picture at left. It is not advisable to attempt drilling the enclosure with a standard bit.

3. **Base Pads** - Peel the backing off of the adhesive base pads (A026) and attach them to the bottom of the enclosure, at the corners. These pads allow the door to swing freely and protects tables from scratches.

4. **Handle and Knob** - Attach the carrying handle (A024) and Door Knob (A025) to the enclosure using the inserted screws.

A024 - Handle

A025 - Knob

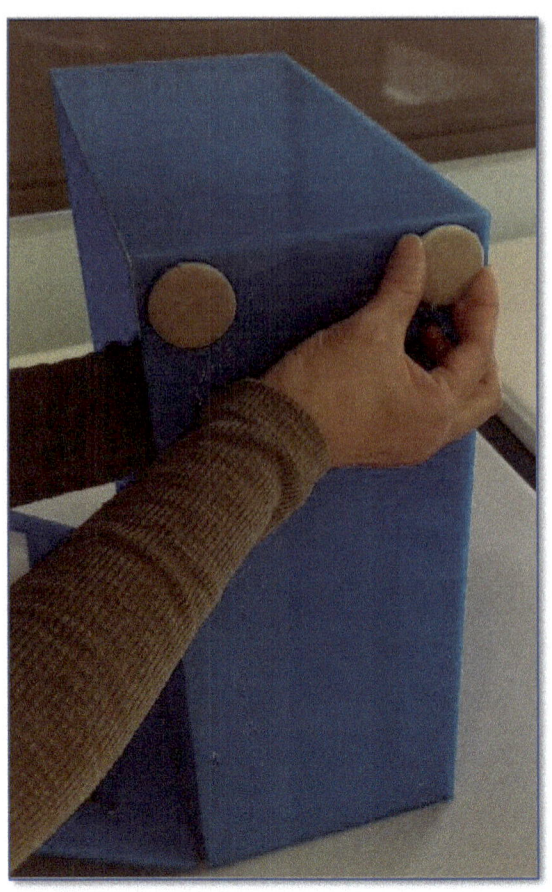

A026 - Pads

5. **Power Button and E-Stop** - Open the boxes containing the Green Lighted Pushbutton (A013) and the Red Twist E-Stop Pushbutton (A014).

Screw the contact blocks into the back side of the buttons. Remove the extra contact blocks (A016) from the bag and attach them also, next to the others. Each button should now have two blocks mounted to the back. When depressing the buttons, you should be able to see the small black plunger move through the hole in the back of the blocks.

The red block has normally closed contacts, while the green blocks have normally open contacts.

A013 – Green PB     A014 – E-Stop PB

Tighten the blocks firmly into the back of the buttons.

A014 – E-Stop PB with 2 contact blocks

By pushing the front of the button into the mounting block and twisting, you should be able to remove the front button part of the two assemblies. Insert them through the top two holes in the enclosure (A001). Before inserting the green button in the top hole, place the Power On Legend Plate (A017) around the button. The E-Stop button goes in the second hole.

There are two angled screws on the back side of the mounting block. After ensuring that the buttons are in the correct position, tighten the screws against the back of the door to secure the buttons in place.

X2   Power, E-Stop and Buzzer

Label (up)

X1

Lamp

Angled Anti-Rotation Screw

A013 – Disassembled Pushbutton

6. Buzzer - Mount the Piezo Buzzer (A015) into the bottom hole. Firmly tighten the collar so that the buzzer doesn't rotate. Ensure that the buzzer label faces toward the top, showing X2 to the left and X1 to the right.

The installed devices should appear as shown in the illustration at left.

Right Side Devices Installed

Power PB

E-Stop

Buzzer

7. Round Buttons - Remove the collar and metal anti-rotation plate from each of the round pushbuttons (A019) and place the buttons into the top row of 8 holes in the device faceplate, from the front. Place the metal anti-rotation plate on the back of the button before screwing the collar back on.

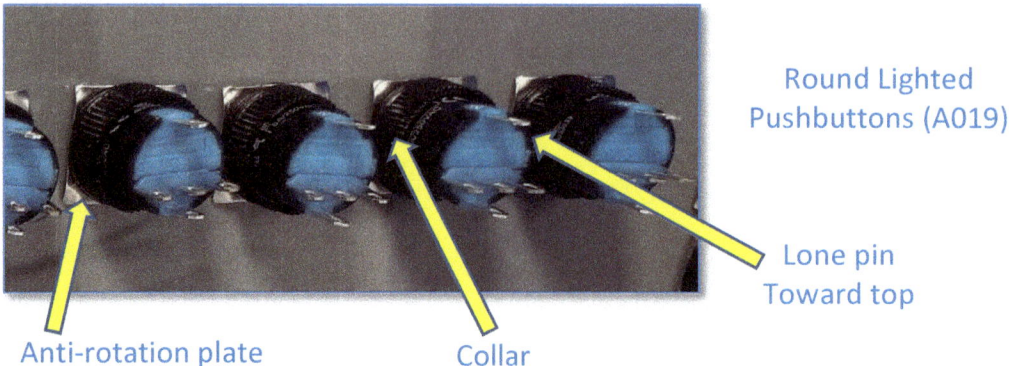

Round Lighted Pushbuttons (A019)

Lone pin Toward top

Anti-rotation plate   Collar

Ensure that the lone pins are toward the top as shown. This is the positive DC terminal for the indicator.

8. Square Indicators - Remove the collar and metal anti-rotation plate from each of the square indicators (A018) and place them into the three holes at the left side of the device faceplate, from the front. Insert the red indicator in the top hole, the yellow indicator in the middle hole, and the green indicator in the bottom hole.

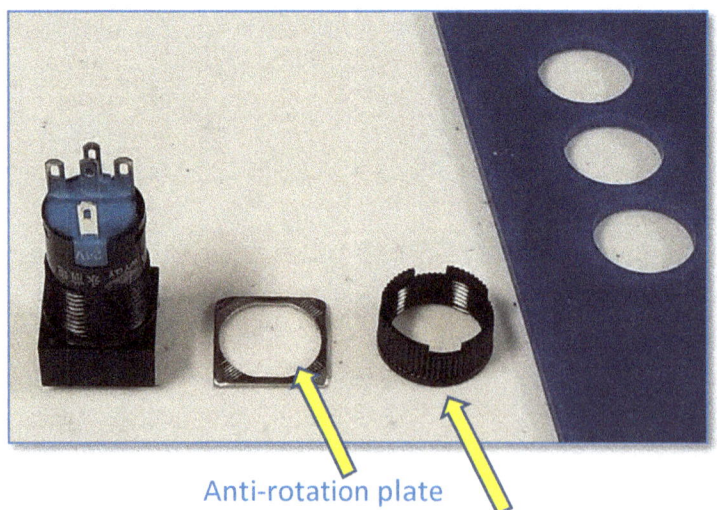

Place the metal anti-rotation plate on the back of the indicator before screwing the collar back on.

These indicators are actually maintained pushbuttons that are used to simulate a stack light.

Anti-rotation plate
Collar

Ensure that the lone pin farthest from the three center pins are oriented toward the outside of the plate as shown below before tightening the collar.

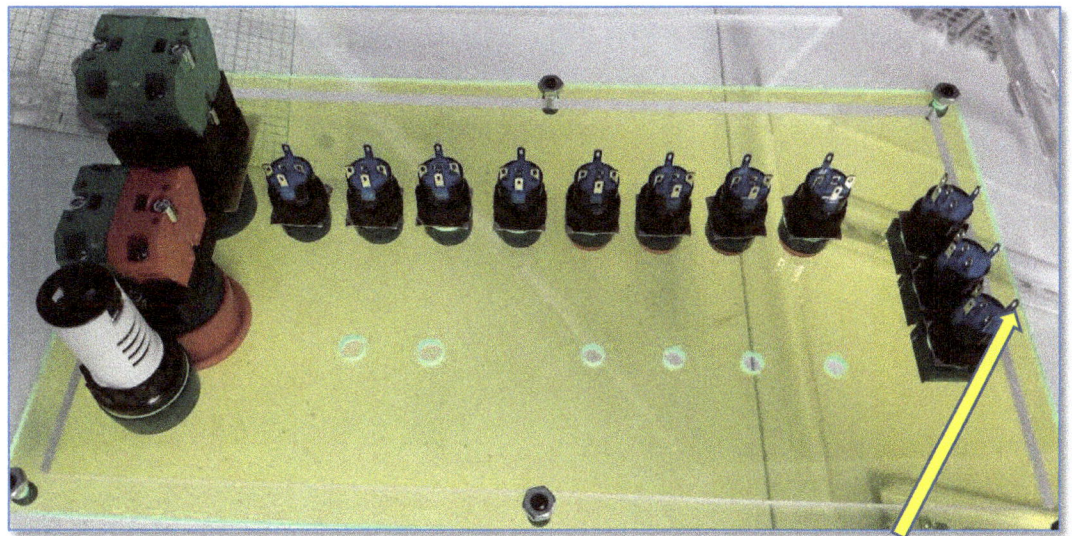

Lone Pins toward Outside

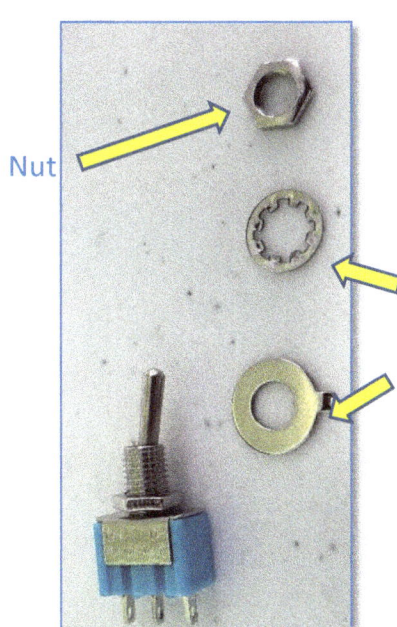

Nut
Lock Washer
Collar

9. Switches - Remove the nut, lock washer and collar from the switches (A021). Place the lock washer back over each switch and insert them through the four holes. Place the collar on the switch from the front and screw the nut onto the switch.

It is optional to remove the bottom nut before installing the switch. The center pin is common, and the bottom pin will connect internally to the common pin when the switch is in the up position.

10. Potentiometers – Remove the washer and nut from each potentiometer (A021) and insert them through the holes from the back. Place the washer over the potentiometer and screw the nut onto the potentiometer.

After tightening the nut with the wires toward the bottom as shown, turn the potentiometer counterclockwise (left) until it stops. Push the knob onto the front of the potentiometer with the pointer in approximately the 7 o'clock position.

 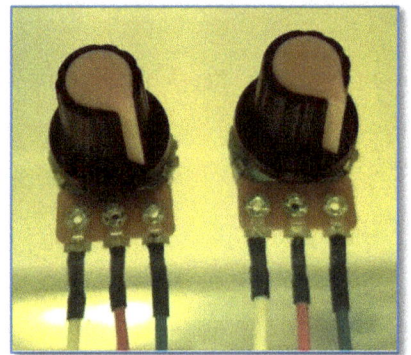

Potentiometer Minimum Position       Potentiometer Maximum Position

After pushing the potentiometer knob onto the shaft, turn it to the minimum and maximum positions. They should move as shown above.

11. **Completed Faceplate Device Mounting** – The completed faceplate should look as shown below. Ensure that all components are tight.

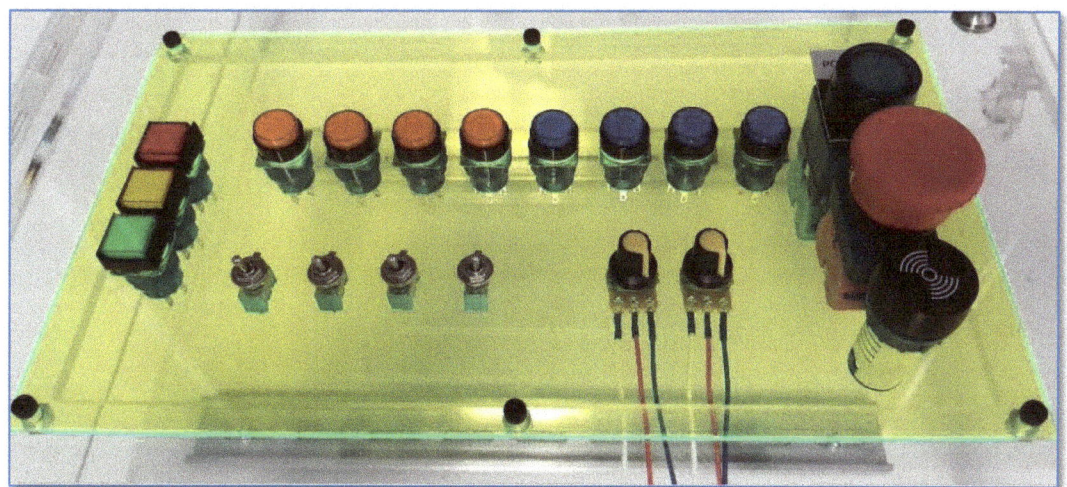

Front of Faceplate

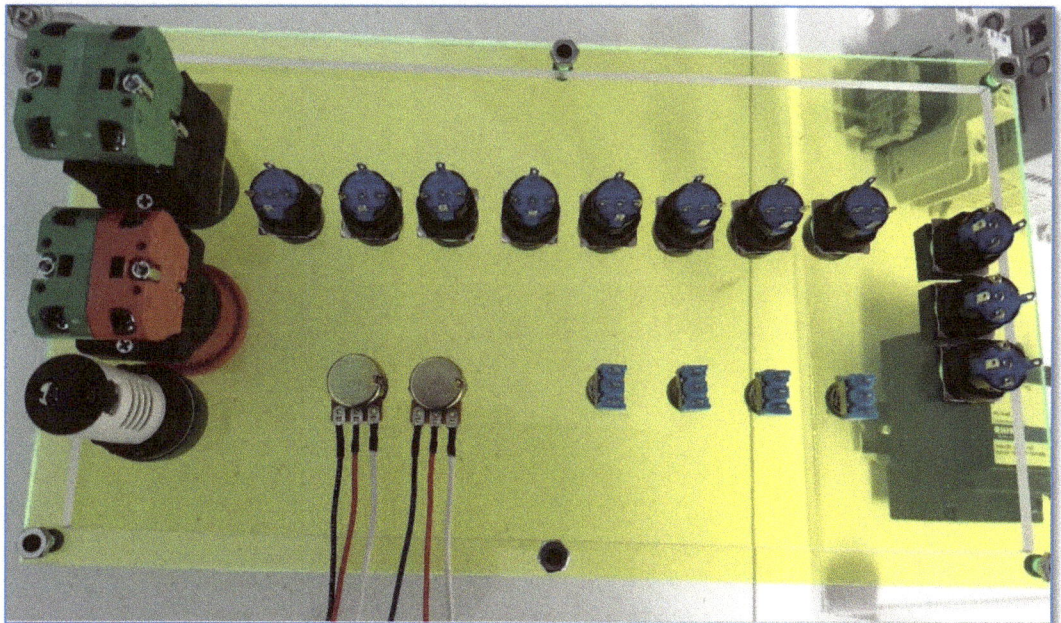

Back of Faceplate

12. End Anchor and Power TBs - Snap an End Anchor (A007) onto the top din rail at the left end. Snap two terminal blocks (A005) next to the end anchor with the exposed metal parts facing to the right as shown.

13. End Cap - Press an end cap (A006) into the side of the terminal block. It should fit firmly into the terminal block holes.

14. Circuit Breaker – Snap the circuit breaker (A004) onto the din rail next to the terminal blocks. Snap another end anchor (A007) next to the breaker as shown.

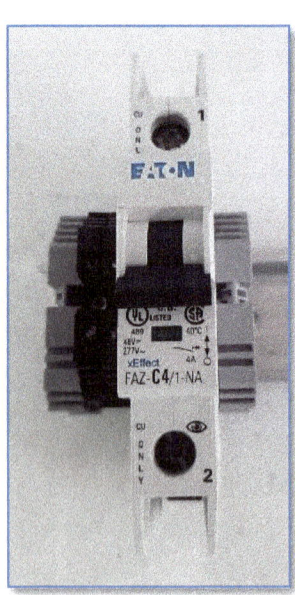

15. PLC – Install the PLC next to the power blocks. The PLC in these pictures is an Allen-Bradley Micrologix 1100; any din-rail mount type PLC can be used. To allow for all of the features of this trainer, it was necessary to add an extra input card and an extra output card to the PLC. These connected to each other with ribbon cables and plugs.

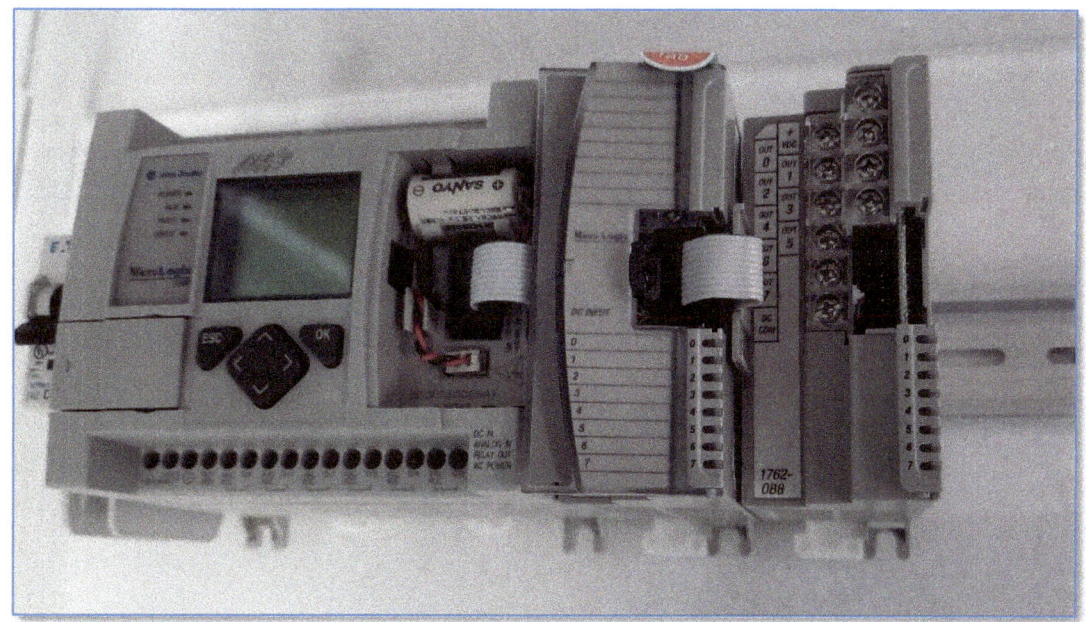

Note the paper tab at the top of the first card, this is an adhesive cover that is used to prevent debris such as metal shavings or stripped wire insulation from falling into the card body. These should not be removed until wiring has been completed. The tabs shown at the bottom of the cards and PLC body can be pulled out before mounting to the din rail and snapped back after installation. Some PLCs may also have tabs at the top of each card that allow the cards to be snapped together from the side using a buss.

If you are building a Standard Trainer or using a non-din-rail mount PLC, you may need to cut the din rail short (for the power blocks) and mount the PLC directly to the back of the trainer. As mentioned before, there are special drill bits used for plastic. A regular tap can be used to tap holes, or a nut can be used on the back of the bolt if using untapped holes. An "acorn nut" can be used for a nice smooth touch.

16. Power Supply – Snap the power supply (A003) onto the bottom din rail at the left side.

17. Fuse Block – Snap the fuse block (A008) onto the din rail next to the power supply. Open it, swinging it down from the top and ensure that the fuse (A009) is inside.

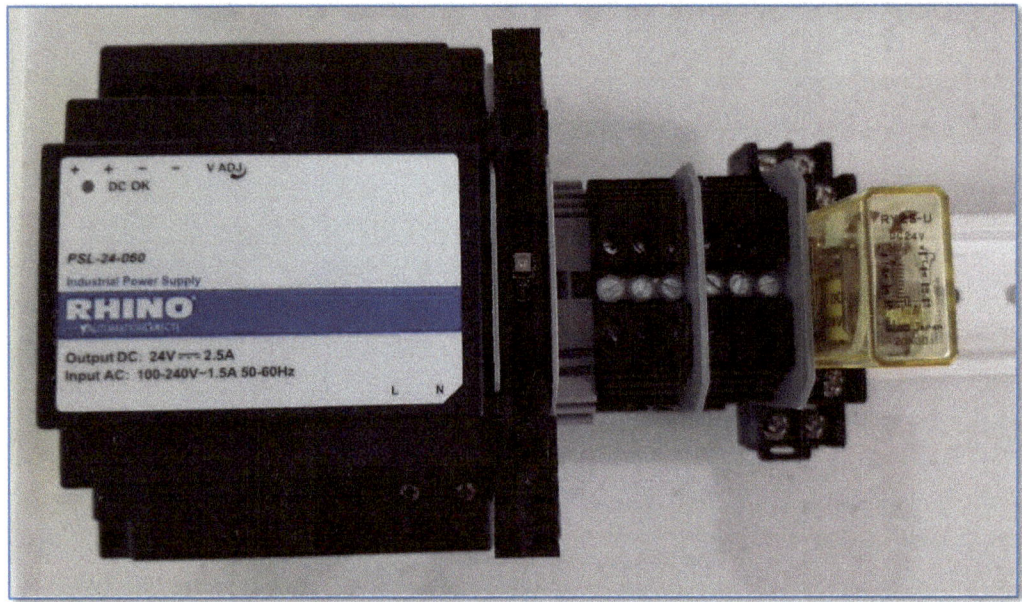

18. DC Terminal Blocks – Insert an end anchor (A007) next to the fuse block. Place three terminal blocks (A005), an end cap (A006), followed by three more terminal blocks and another end cap, as shown. Drop a set of jumpers (A010) into the center of the terminal blocks and tighten the screws.

19. Relay and Socket - Snap the relay socket (A023) onto the din rail and insert the relay (A022).

20. Wire Duct – Remove the adhesive backing from the tape on the back of the wireway (A011) and stick it centered underneath the terminals and cards of the PLC.

## Wiring

Prepare your tools and wire for the wiring of the trainer. There are three bundles of wire as listed in the beginning section of this guide. One bundle has spiral wrap twisted around it, this bundle will be used last.

As you strip wires before terminating them, small pieces of insulation will be created that you will need to collect and dispose of. As mentioned previously, ferrules are optional; wires can be twisted at the ends and terminated without them.

### PLC I/O List and Wiring Notes

The PLC used in these pictures and examples is an Allen-Bradley Micrologix 1100 as mentioned in the Introduction of this document. Depending on the type of PLC used, your connections will be different, but the concept of where to connect things is the same.

**DC Common (DC COM)** terminals are associated with inputs. For positive inputs, they are wired to the -DC terminal blocks.

**VDC** terminals are wired to the positive power source, and may be switched or unswitched. Moving actuators are usually switched, that is, they are disconnected by the MCR when the EStop button is pressed. The pilot lights or indicators behind the pushbuttons are wired this way, the E-Stop disconnects the power feed from the lights.

The **Unswitched Outputs** are used for the buzzer and the simulated stack light.

| Inputs | Devices | Notes |
|---|---|---|
| I:0/0-I:0/7 | Pushbuttons 1-8 | Lighted buttons on door (A019) |
| I:0/8-I:0/9 | Switches 1-2 | Toggle switches on door (A020) |
| I:1/0-I:1/1 | Switches 3-4 | Toggle switches on door (A020) |
| I:1/2 | E-Stop | Red maintained pushbutton on door (A014) |
| I:1/3 | Reset | Green lighted button on door (A013) |
| I:1/4 | MCR | Master Control Relay in enclosure (A022/A023) |
| I:1/5-I:1/7 | Spare | Not used |
| **Outputs** | **Devices** | **Notes** |
| O:0/0 | Red Stack Light | Red Light indicator, left side of door (A018) (Unsw.) |
| O:0/1 | Yellow Stack Light | Yellow Light indicator, left side of door (A018) (Unsw.) |
| O:0/2 | Green Stack Light | Green Light indicator, left side of door (A018) (Unsw.) |
| O:0/3 | Buzzer | Piezo Buzzer on Door (A015) (Unsw.) |
| O:0/4-5 | Spare | Not used, wired as switched |
| O:2/0-O:2/7 | Indicators 1-8 | Lighted buttons on door (A019) (Switched) |
| **Analog In** | **Devices** | **Notes** |
| IV1, IV2 | Potentiometers | 1K Ohm, 0-10VDC |

If you are using a different PLC, you will want to make an I/O list of your own. The only restrictions are which outputs can be used as **switched** or **unswitched**, based on which VDC terminals feed which outputs.

21. AC Wiring – Using the red wire from the loose power wiring bundle (A029), measure and cut it to the proper length, strip both ends, and connect the bottom of the circuit breaker (A004) to the terminal labeled "L" on the power supply (A003). Measure, cut, strip and connect the white wire from the bundle from the bottom of the first terminal block (A005) next to the circuit breaker to the terminal labeled "N" on the power supply.

    Using the green wire from the loose power wiring bundle (A029), connect the terminal block next to the one you connected the white wire to, to the second set of jumpered terminal blocks next to the relay. These are the -DC (Common) blocks.

    If your PLC is AC powered: Connect a red wire from the circuit breaker (A004) to the power terminal of your PLC. This will usually be labeled L or L1. Connect a white wire from the terminal block you previously connected to the power supplies' neutral terminal to the neutral terminal of your PLC. This will usually be labeled N or L2. Connect a green wire from the ground terminal block (next to the neutral) to the ground terminal of the PLC.

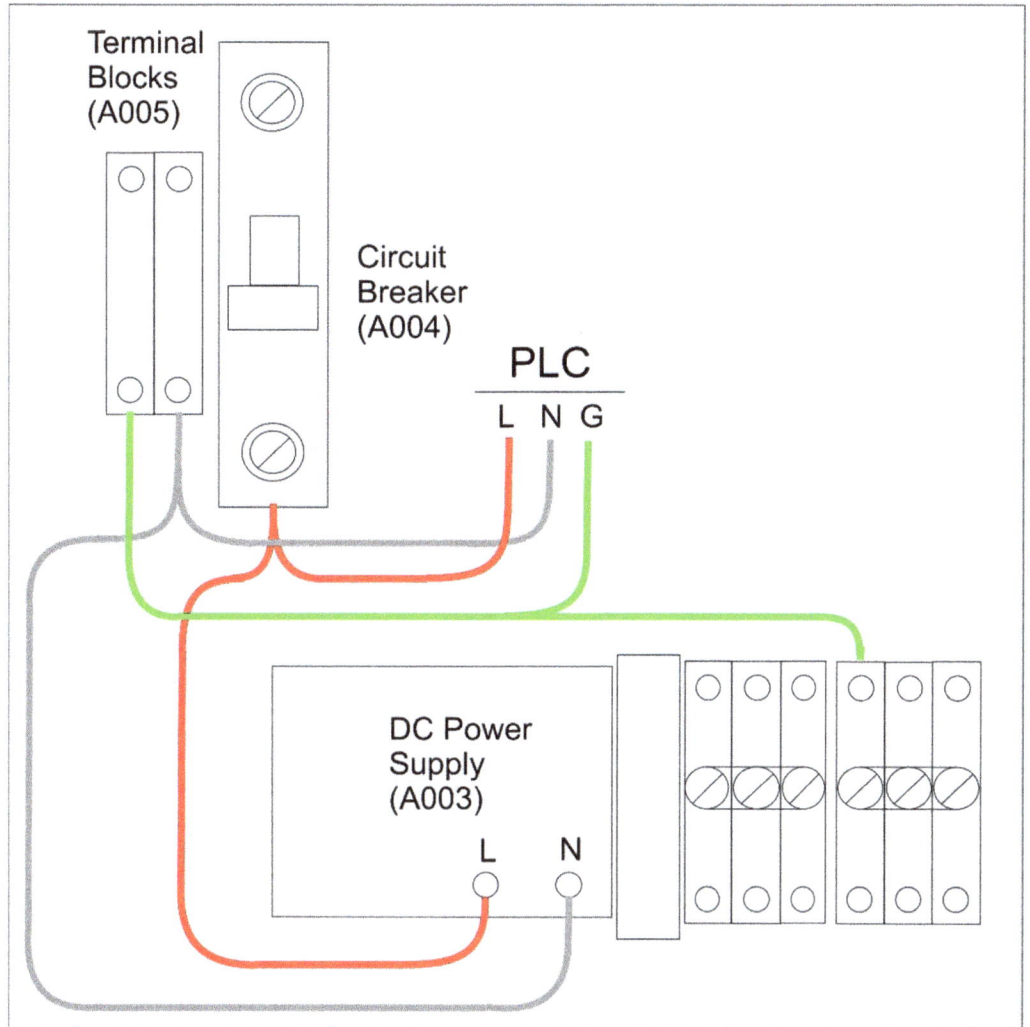

AC Wiring Step 21

22. DC Wiring (a) – Using the <u>blue wire</u> from the loose power wiring bundle (A029), connect the power supply (A003) "+" terminal to the top of the fuse block (A008). Connect the bottom of the fuse block to a terminal block in the group of three adjacent to the fuse block. These are the +DC supply terminals. From the bottom of the +DC terminal blocks, connect to terminals 12 and 9 of the relay socket (A022). It may be easier to remove the relay from the socket.

Using the <u>blue/white stripe wire</u>, connect the power supply (A003) "-" terminal to the top of the second group of three terminal blocks. These are the -DC supply terminals. From the bottom of the -DC terminal blocks, connect to terminal 13 of the relay socket (A022). This is the negative terminal of the relay coil.

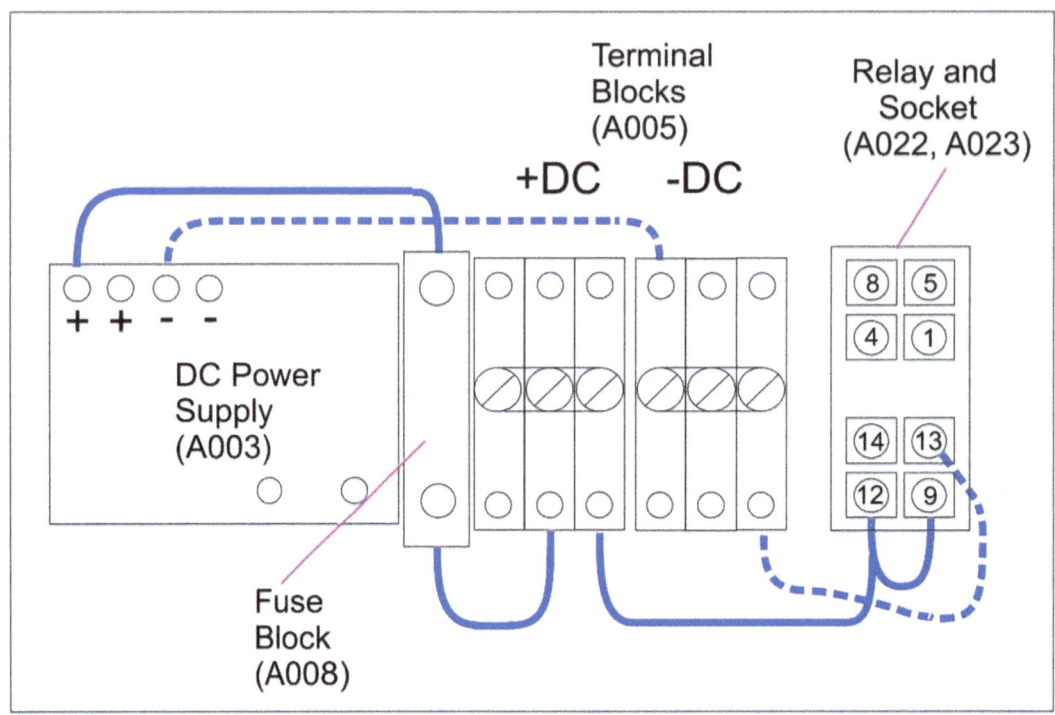

DC Wiring Step 22

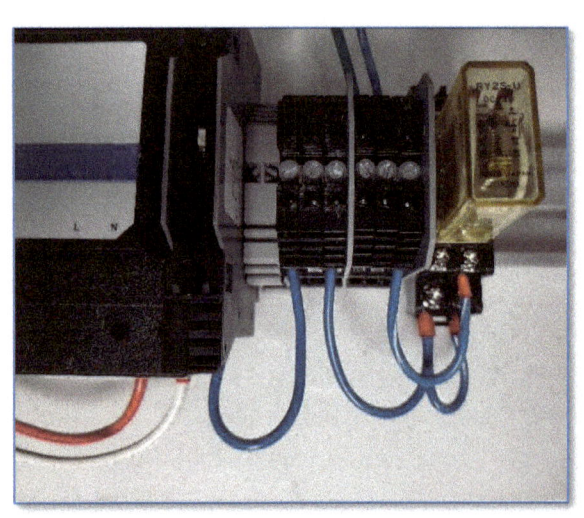
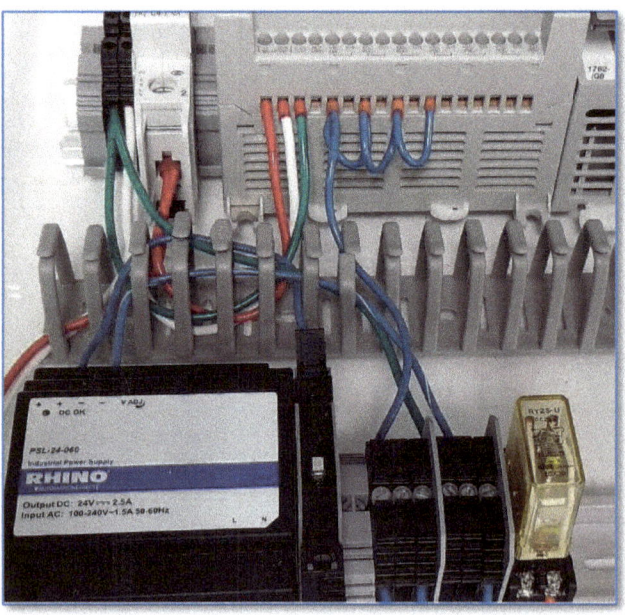

AC and DC wiring, steps 21 and 22

23. DC Wiring (b), Output Power – Using blue wire from the loose power wiring bundle (A029), connect from the +DC terminal blocks to the unswitched power feeds (VDC) for the DC outputs of the PLC.

    Using orange wire from the loose control wiring bundle (A028), wire from pin 5 of the relay socket (A022) to all of the switched power feeds (VDC) for the outputs. Finally, from the last switched VDC terminal, connect to the input assigned to the MCR.

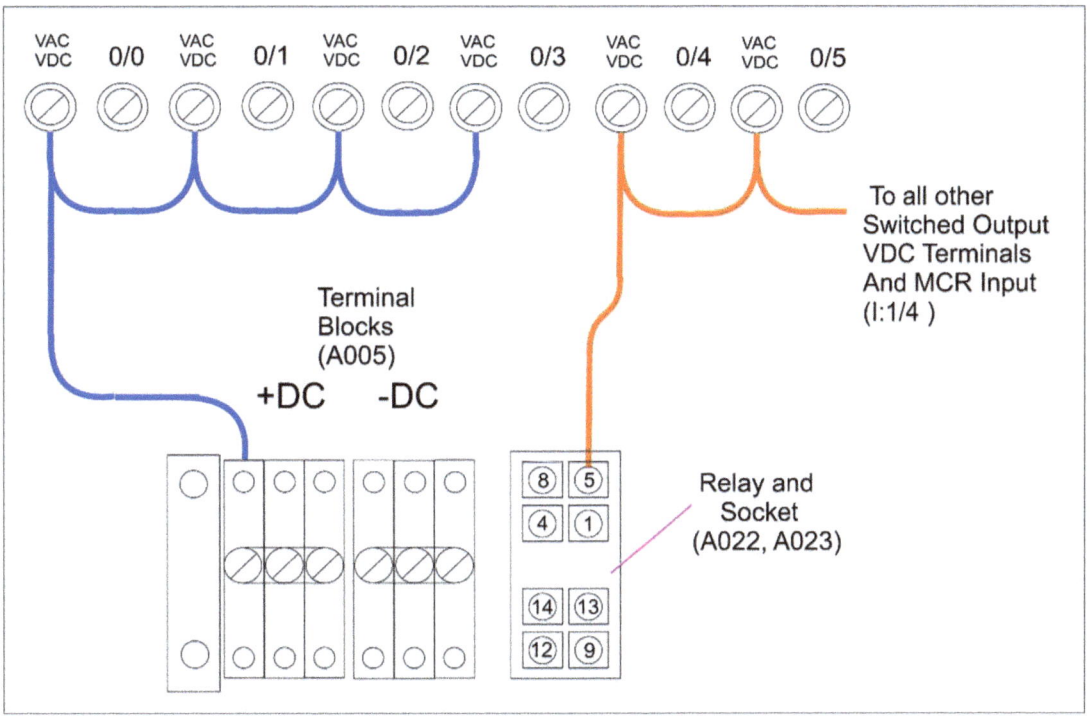

DC Switched and Unswitched Output Power wiring, step 23

24. DC Wiring (c), Input/Output/Analog Commons – Using blue/white stripe wire from the loose power wiring bundle (A029), connect from the -DC terminal blocks to all of the DC COM terminals for the inputs and outputs. Also connect to the COM or – terminal of the analog inputs.

# Building and Wiring Your Trainer Kit

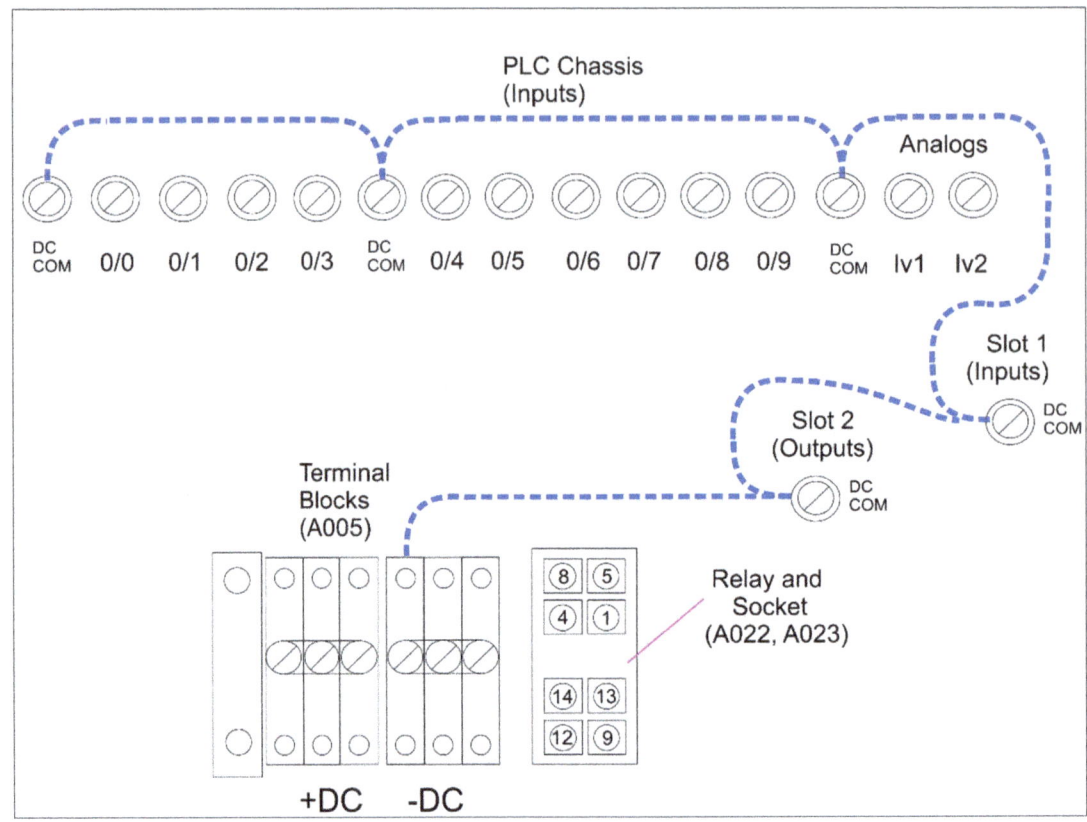

-DC Common Power wiring, step 24

After completing steps 21 through 24, the wiring might look something like the picture below:

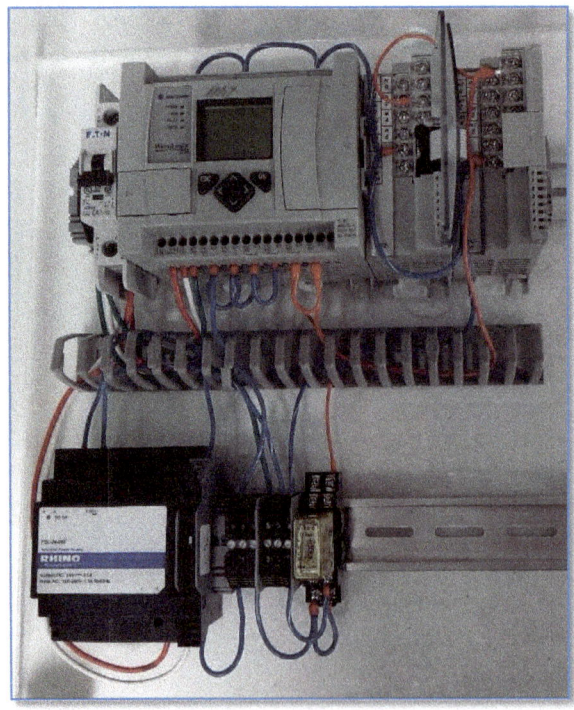

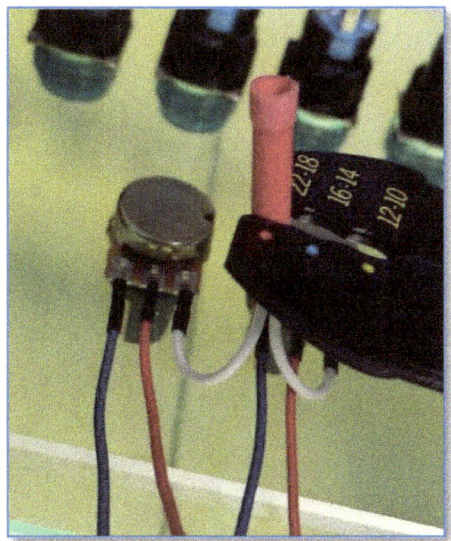 

*Crimped potentiometer butt splices, step 25*

25. Potentiometer Butt Splices - The analog potentiometers (A021) are connected to the DC power supply and analog I/O points by means of crimped butt splices (A032). Since the positive and negative power feeds will be connected next, it is easiest to do all of the potentiometer connections together. Strip the ends of the two white wires about ½" and twist them together. Push the twisted ends into one side of the but splice and firmly crimp the connector onto the wires. Do the same with the two blue wires, connecting them together. Crimp one butt splice firmly onto each red wire. After completion check to ensure that the splice is tightly crimped to the wires.

26. DC Wiring, Door -DC connections – Using gray wire from the loose control wiring bundle (A028), strip and crimp one end to the butt splice on the two white wires of the potentiometers (A021). You will be making a "daisy chain" of connections, feeding the -DC to the devices mounted to the door. Measure the distance of the wire to the terminal on the buzzer (A015), cut and strip the wire to the correct length so that the end fits easily under the terminal. This should be approximately 2.5". Cut and strip the end of the remaining gray wire and twist the ends together. If using ferrules, the ferrule can be crimped over the wires. **Note below that component numbers appear reversed from the back!**

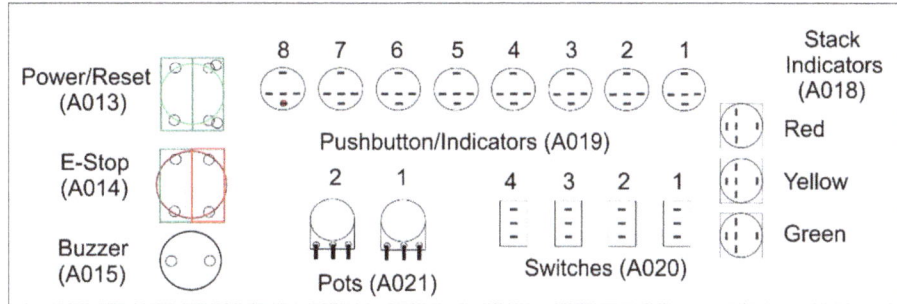

**Tip:** You may find it easier to connect all of the wires together as you measure, cut and strip them, before pushing them onto the pins, as shown below.

Remove the contact blocks from the back of the Power/Reset button (A013). Measure, cut and strip the gray wire so that it reaches the bottom terminal of the button. This is the DC – connection for the lamp. Again, strip the insulation from the end of the remaining gray wire, twist the wires together, and if using a ferrule you may place it over the twisted wires and crimp it. If not using ferrules it may be easier to tighten the twisted wires under the buzzer and pushbutton to hold them in place.

Measure, cut and strip the remaining gray wire so that it reaches the bottom terminal of the closest pushbutton/indicator (A019). Again, measure, cut and strip the remaining gray wire and twist the ends together. Insert the ends into the **Uninsulated** 0.108 F spade connectors (A031) and crimp them tightly onto the twisted ends.

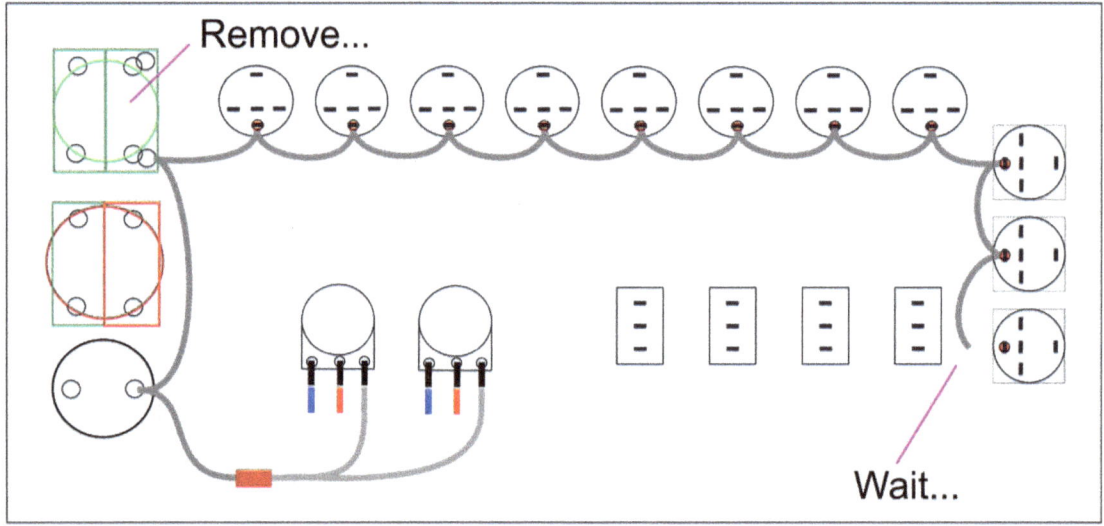

-DC door connections, step 26

The diagram above shows all of the connections you will be making with the gray wire. You may find it easier to cut and strip all of the wires that run between each of the lighted pushbuttons at the top, they are all about 2.5" in length. Continue cutting, stripping and crimping the spade connectors onto the wires. Between the last pushbutton (A019) and the top red square indicator (A018), the length may be slightly different. Wait! Do Not Crimp the Last Terminal!

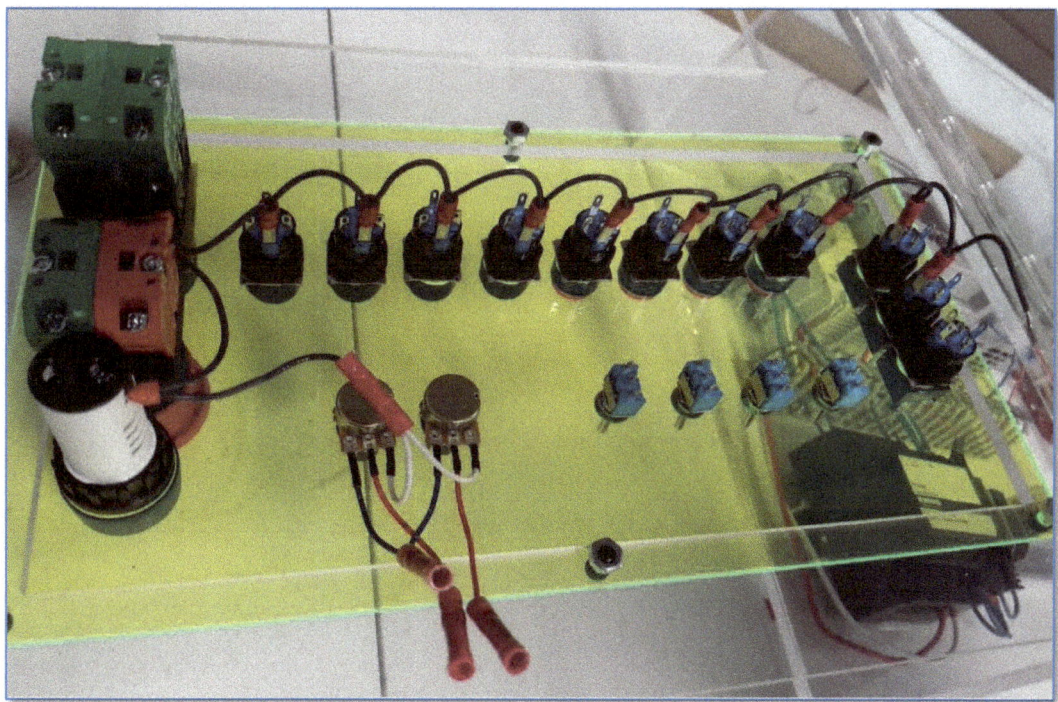

DC- door connections (completed), step 26

After completion of the wiring connections for the -DC, push the spade terminals onto the pins on the buttons as shown. Notice that the last wire is not crimped and connected yet, it will be done later.

27. DC Wiring, Door +DC connections – Using brown wire from the loose control wiring bundle (A028), strip and crimp one end to an **Uninsulated** 0.108 F spade connector (A031). This can be pushed onto the center pin of the farthest right switch (A020). As with the gray wires, you will be "daisy-chaining" the wires together to provide the positive DC power to all of the door devices. As with the previous DC wiring, the distance between switches and pushbuttons for wires is about 2.5". Connect all of the center pins on the switches. **Leave them very loose! They will need to be removed!**

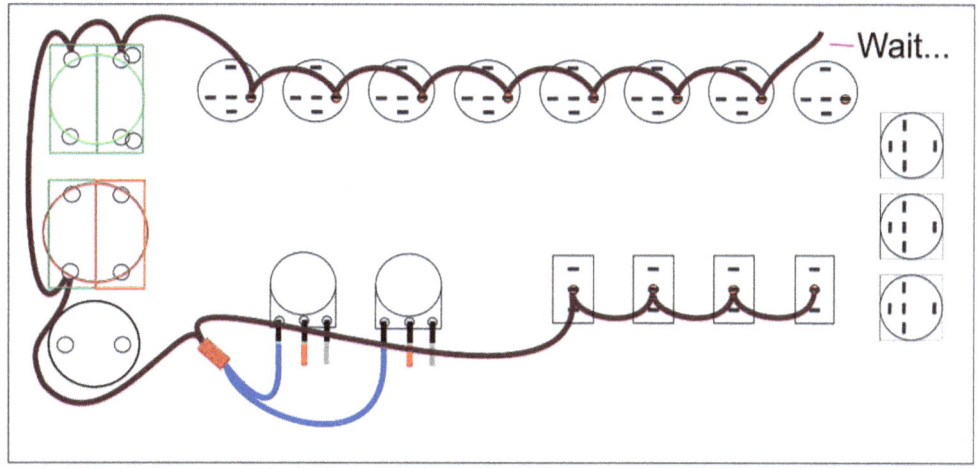

Connect the center pin on the last switch using a spade terminal crimp to the butt splice connecting the two blue wires on the potentiometers (A021). Connect the Potentiometer to the bottom of the green contact block on the E-Stop pushbutton (A014). Wires can be twisted together here or a ferrule can be used. Connect a brown wire from here to the tops of the two contact blocks on the Power/Reset button (A013), again using ferrules if desired.

From the Power/Reset contact blocks, connect to the farthest right terminal on the closest pushbutton/indicator (A019). Jumper to each of the far-right terminals as shown in the diagram, using **Uninsulated** 0.108 F spade connectors (A031). As with the previous -DC wiring chain, Wait! Do Not Crimp the Last Terminal!

After completion of the brown +DC wiring, the door should look as shown below:

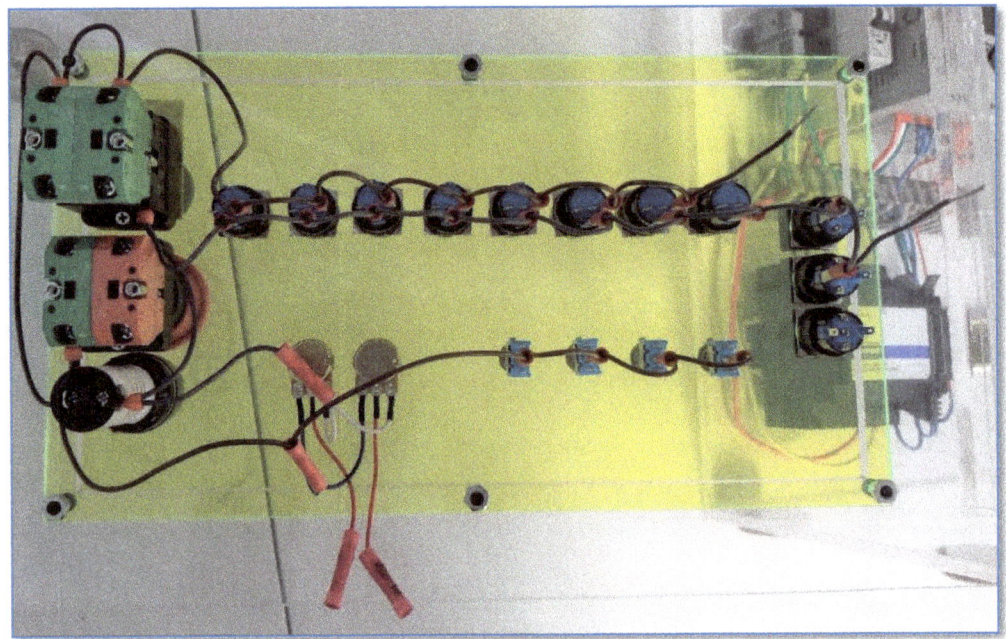

DC + and - Door Connections (completed), step 27

28. DC Extension to Terminal Blocks – Using the blue wire from the from the loose control wiring bundle (A028), twist it together with the gray wire that you left loose from the square indicator (A018). Crimp an uninsulated 0.108 F spade connector (A032) over the end and attach it to the left terminal of the green stack light indicator. Using a brown wire from the loose control wiring bundle (A028), twist it together with the brown wire that you left loose from the last pushbutton (A019). Crimp an uninsulated 0.108 F spade connector (A032) over the end and attach it to the right terminal of the last pushbutton.

Important Note: **Descriptions in the following steps describe connecting wires to the I/O points on the PLC for simplicity. You may want to wait until all door connections have been made and the harnessing is dressed before making these terminations!**

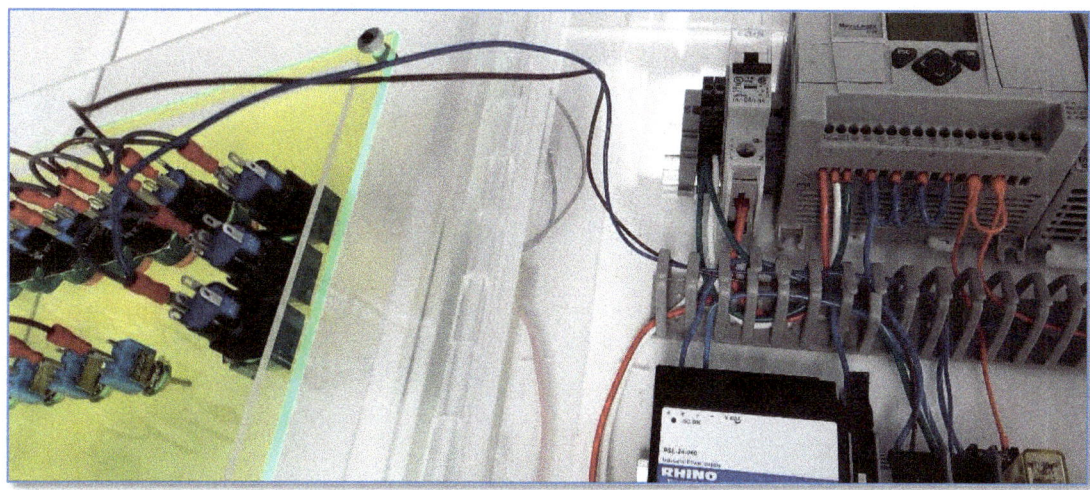

*DC + and - Door Connections (completed), step 28*

Run both wires through the wireway (A011), dropping them through the slots into the DC terminal blocks. Connect the blue wire to a -DC terminal, and the brown wire to a +DC terminal. Leave enough slack in the wire so that it doesn't touch the door hinge with it open.

29. Door to MCR Wiring - Using the <u>white wire</u> from the from the loose control wiring bundle (A028), connect the top of the base on the power/reset button (A013) to the top of the red contact block of the E-Stop button (A014). Connect the remaining white wire to the same point (twist wires together or use ferrule). You will probably need to remove the red contact block to connect the wires.

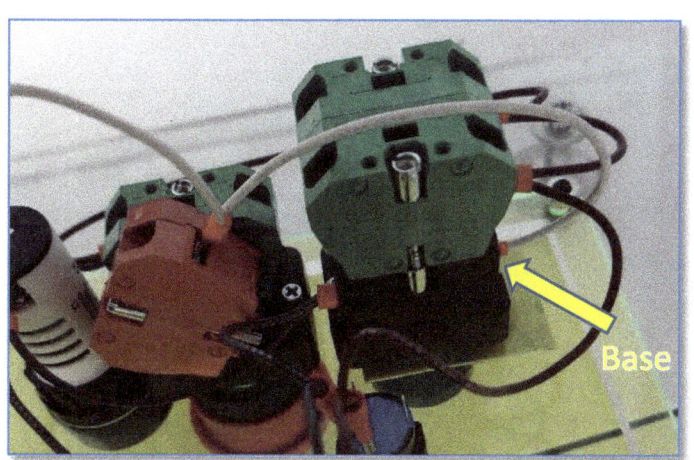

Using the <u>purple wire</u> from the loose control wiring bundle (A028), connect the bottom of the left side contact block of the power/reset button (A013) to the bottom of the red contact block of the E-Stop button (A014). Connect the remaining purple wire to the same point (twist wires together or use ferrule).

Run the white and purple wires along the bottom part of the wiring on the face plate, along the line of the potentiometers and switches. The wires should meet up with the brown and blue wires you ran previously, you may find it convenient to use a twist tie to hold the wires together temporarily, do not use tie wraps yet, or if you do, do not pull them tight.

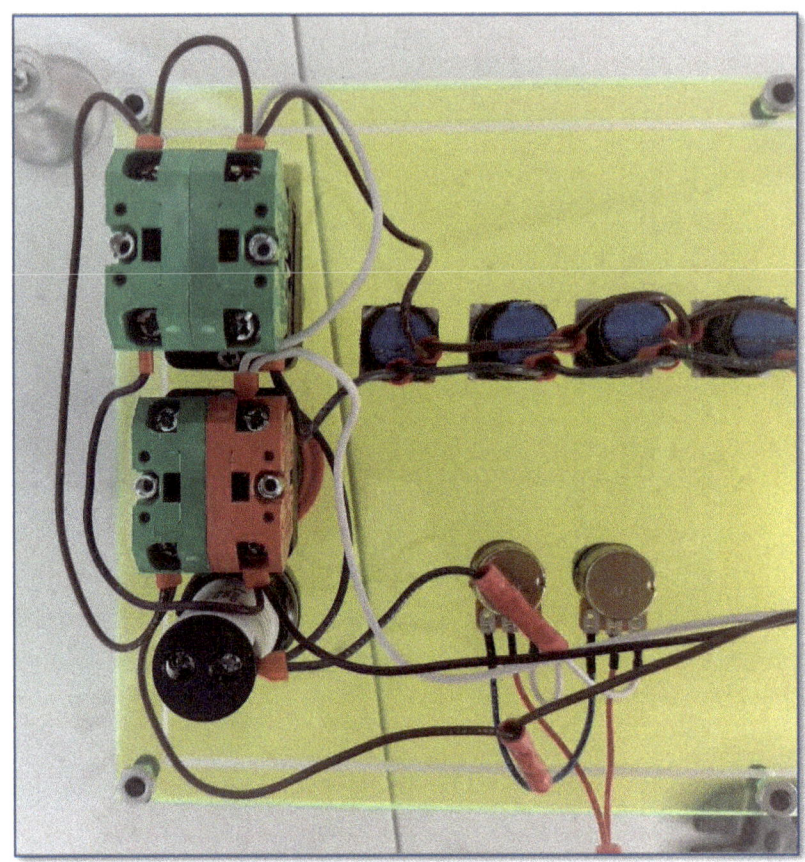

Door Connections for MCR, step 29

Run both wires through the wireway (A011), dropping them through the slots toward the MCR relay (A022, A023). Connect the white wire to pin 14, and the purple wire to pin 8. Route the wires alongside the blue and brown wires, leaving the same amount of slack.

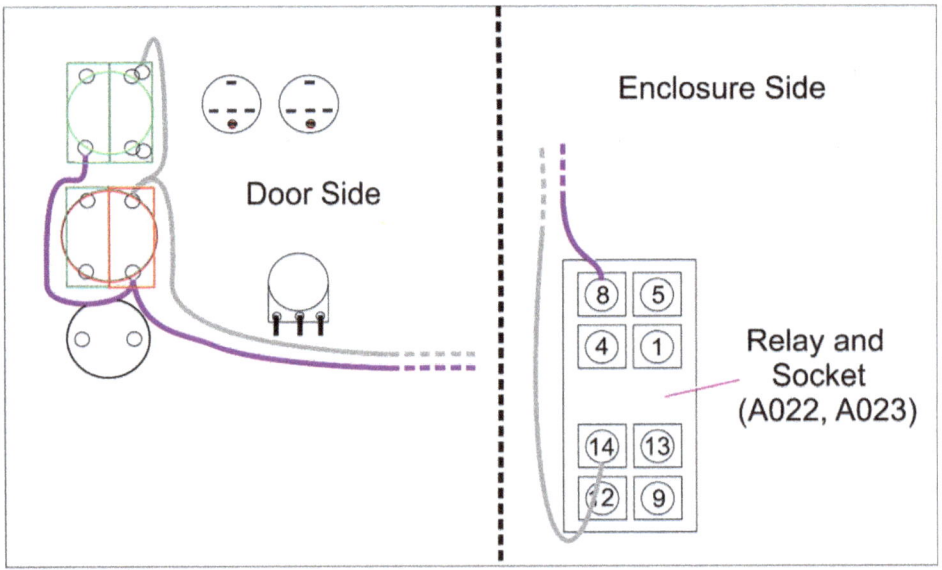

30. Door Power/E-Stop I/O Wiring – Using the white/violet stripe wire from the device cable with spiral wrap bundle (A027), connect to the bottom of the right-hand contact block of the power/reset button (A013). Connect the other end to your assigned Reset input on the PLC (I:1/3). Using the white/brown stripe wire from the same bundle, connect to the top of the green contact block of the E-Stop button (A014). Connect the other end to your assigned E-Stop input of the PLC (I:1/2). Run the wires along the line of the top row of pushbuttons to the door hinge, meeting the other group of wires that go to the backplane. Again, you may find it convenient to temporarily wrap the wires together with a twist-tie.

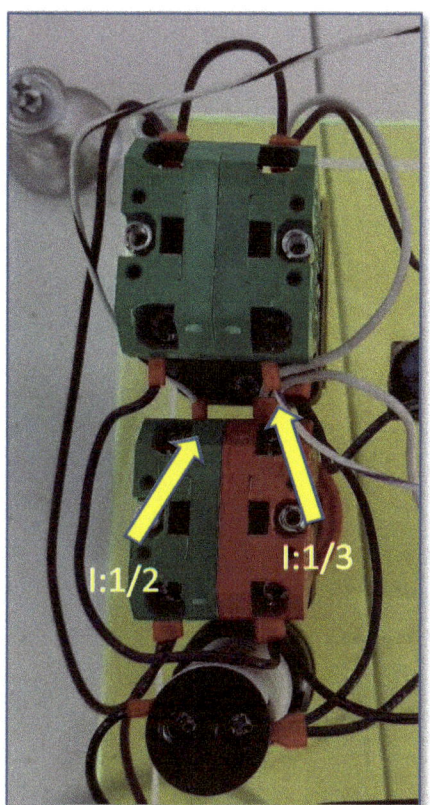

Power and Estop I/O, step 30

Temporary twist-tie

31. Stack Light Indicators I/O Wiring – Using the colors listed below from the loose control wiring bundle (A028), connect the outer pins of the square indicators (A018) to the listed I/O point. Use the **Uninsulated** 0.108 F spade connectors (A031) on the indicator pin.

| Device | Output | Wire Color |
| --- | --- | --- |
| Red Light (Top) | O:0/0 | Red |
| Yellow Light (Middle) | O:0/1 | Yellow |
| Green Light (Bottom) | O:0/2 | Green |

Stack Light I/O, step 31

**Important Note:** The door connections can all be made and dressed before the other ends of the wires are attached inside the enclosure. This may make length measurements and wire dressing easier.

32. Buzzer I/O Wiring – Using the black wire from the loose control wiring bundle (A028), connect the remaining outside terminal of the piezoelectric buzzer to your assigned buzzer output on the PLC (O:0/3).

33. Switch I/O Wiring - Using the colors listed below from the device cable with spiral wrap bundle (A027), connect the bottom pins of the switches (A020) to the listed input point. Use the **Insulated** 0.108 F spade connectors (A030) on the switch pin.

    | Device   | Input | Wire Color/Stripe |
    |----------|-------|-------------------|
    | Switch 1 | I:0/8 | White/Red         |
    | Switch 2 | I:0/9 | White/Orange      |
    | Switch 3 | I:1/0 | White/Yellow      |
    | Switch 4 | I:1/1 | White/Green       |

34. Potentiometer I/O Wiring – Using the white/blue stripe wire from the device cable with spiral wrap bundle (A027), crimp it firmly to the red wire butt splice of the first potentiometer (closest to the switches). Connect the other end to your assigned potentiometer 1 analog input (IV1). Using the white/black and red stripe wire from the same bundle, crimp it to the other potentiometer butt splice. Connect the other end to your assigned potentiometer 2 analog input (IV2).

    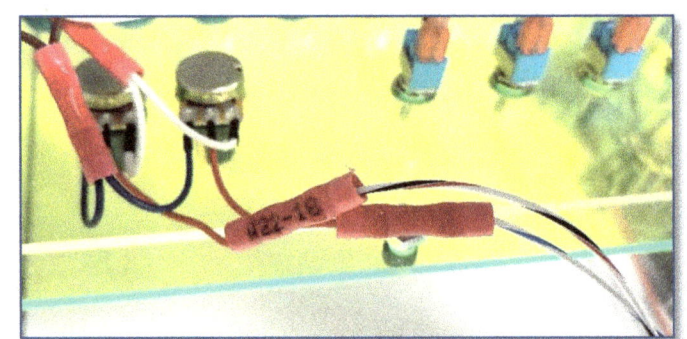

    Potentiometer I/O, step 34

35. Pushbutton Input Wiring – Using the colors listed below from the device cable with spiral wrap bundle (A027), connect the middle pin of the pushbuttons (A019) to the listed input point. Use the **Insulated** 0.108 F spade connectors (A030) on the pushbutton pins.

    | Device   | Input | Wire Color/Stripe |
    |----------|-------|-------------------|
    | Button 1 | I:0/0 | Black             |
    | Button 2 | I:0/1 | Brown             |
    | Button 3 | I:0/2 | Red               |
    | Button 4 | I:0/3 | Orange            |
    | Button 5 | I:0/4 | Yellow            |
    | Button 6 | I:0/5 | Green             |
    | Button 7 | I:0/6 | Blue              |
    | Button 8 | I:0/7 | Violet            |

It can be difficult to estimate the correct length to cut the wires as they are run to the inputs and outputs of the PLC. Using twist ties or loosely wrapping the tie-wraps (A033) around the wires can help. This is why you may want to make I/O connections later.

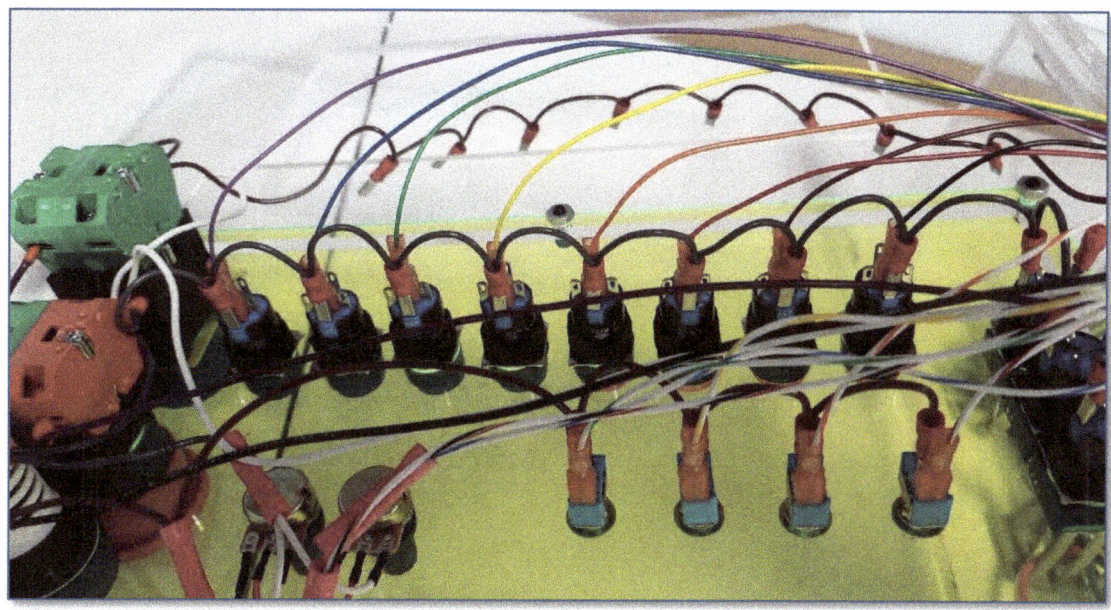

Pushbutton Inputs, Step 35.

Also, the brown jumpers previously connected to the outside pins will have to be removed as shown to fit the insulated spade terminals onto the pins. This is why Step 27 said to leave them loose!

36. Indicator Output Wiring - Using the colors listed below from the device cable with spiral wrap bundle (A027), connect the top pin of the pushbutton/indicators (A019) to the listed input point. Use the **Uninsulated** 0.108 F spade connectors (A031) on the indicator pins.

| Device | Output | Wire Color/Stripe |
|---|---|---|
| Indicator 1 | O:2/0 | Gray |
| Indicator 2 | O:2/1 | White |
| Indicator 3 | O:2/2 | Pink |
| Indicator 4 | O:2/3 | Tan |
| Indicator 5 | O:2/4 | Red/Black |
| Indicator 6 | O:2/5 | Red/Yellow |
| Indicator 7 | O:2/6 | Red/Green |
| Indicator 8 | O:2/7 | White/Black |

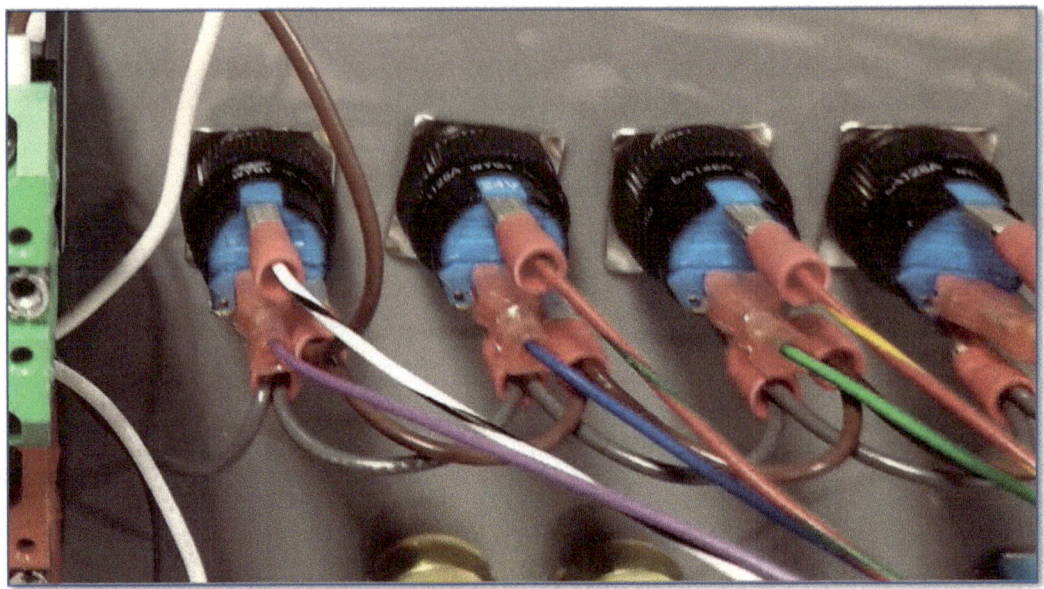

Pushbutton/Indicator Pin Configuration, Step 36.

After completion of step 36, the pushbutton/indicators (A019) should appear as shown above. To insert the insulated spade terminals (A030), the uninsulated jumpers next to them had to be removed first, then re-inserted afterwards.

37. Door Tie-Wrapping/Harness Dressing – At this point all of the wires from the door to the inside of the enclosure have been installed, and most of the wiring is complete. You should have been able to manage the wire lengths so that they are relatively uniform, without too much extra wire (which you would have to fold into the bundle) or too little (making the bundle too tight against the hinge).

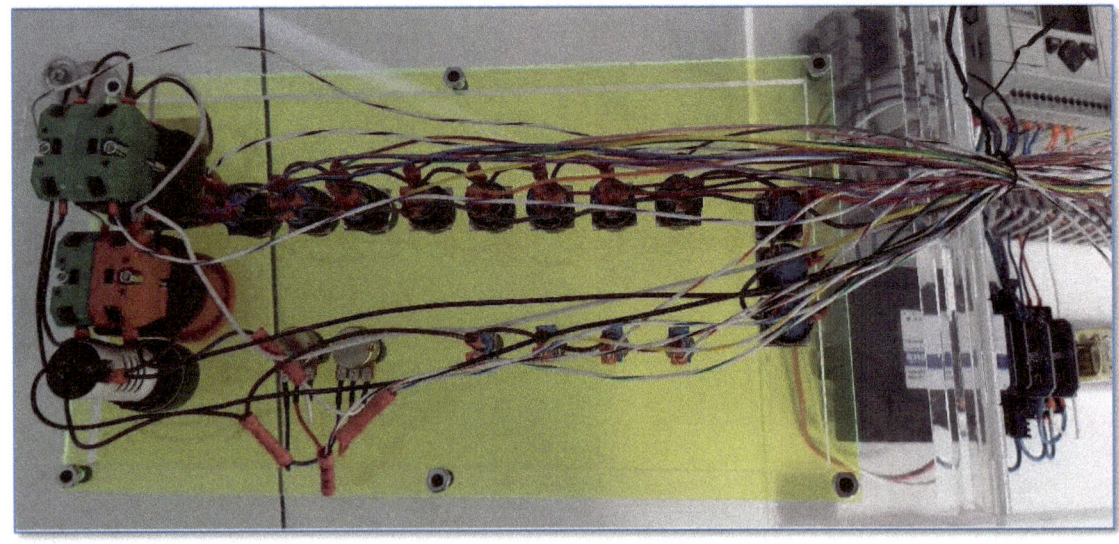

Before dressing the wires, Step 37.

Starting from the outside edge of the door, use the tie-wraps (A033) to lace the various wires together neatly. There can be a bit of an art to this, and there are only so many tie wraps…

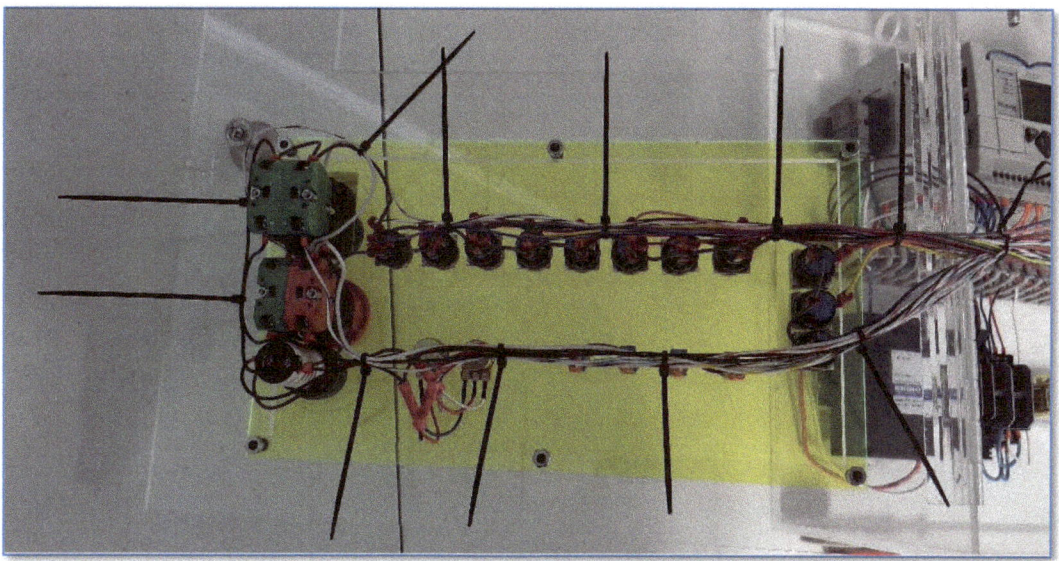

Partially dressed, Step 37.

The picture above shows what the door might look like after using some of the tie wraps. It is important not to tighten the tie wraps completely until you have pulled the wires completely into the desired location, and not to cut the tie wraps protruding end until it is tightened.

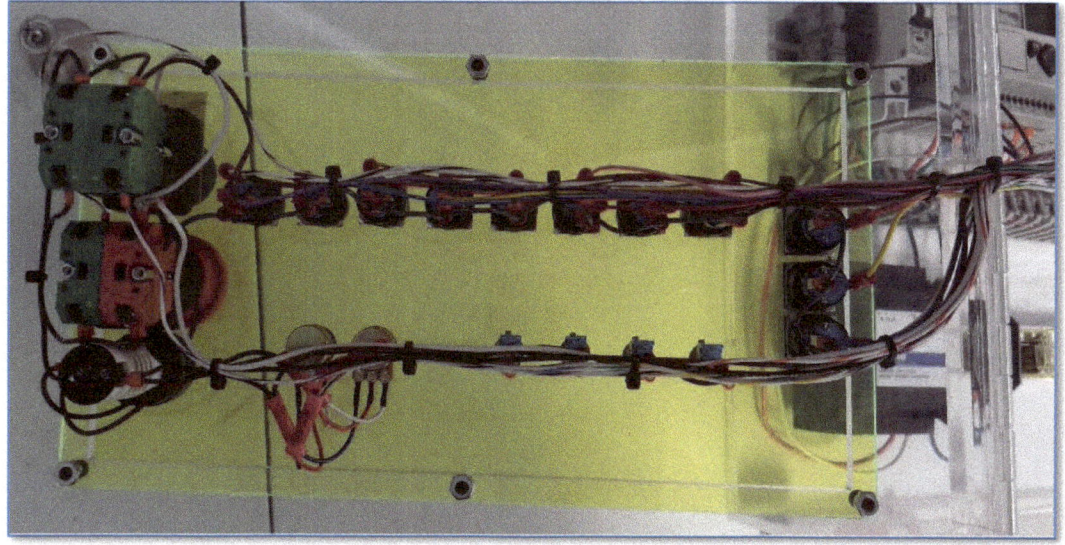

Door Dressed and Completed, Step 37.

After completing the door, you can tighten and cut the protruding part of the tie-wrap flush with the eyelet.

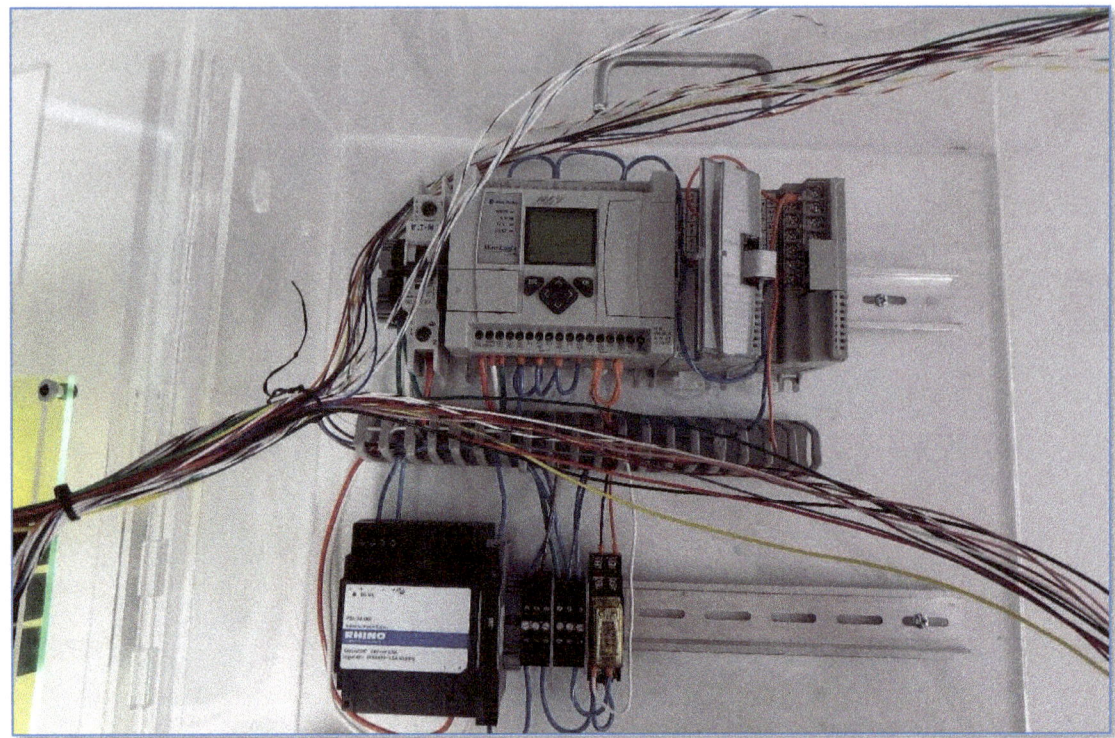

Before Wiring I/O, Step 37.

The above picture shows the wires from the door before being terminated into the PLC I/O points. In this example the MCR and DC door wiring was terminated early, but the I/O wiring was done after that. As mentioned previously, this can make wire length estimations and dressing easier. The wires have been separated into groups based on whether they will be terminated into the top or the bottom of the PLC.

Bottom partially wired, Step 38.

38. I/O Terminations – As mentioned in the previous notes, it may be easier to terminate I/O points after completing the door wiring. I/O terminations were listed in the previous steps for simplicity.

It is also easier to connect the bottom row of multi-level terminal arrangements as shown below, then connect the upper level!

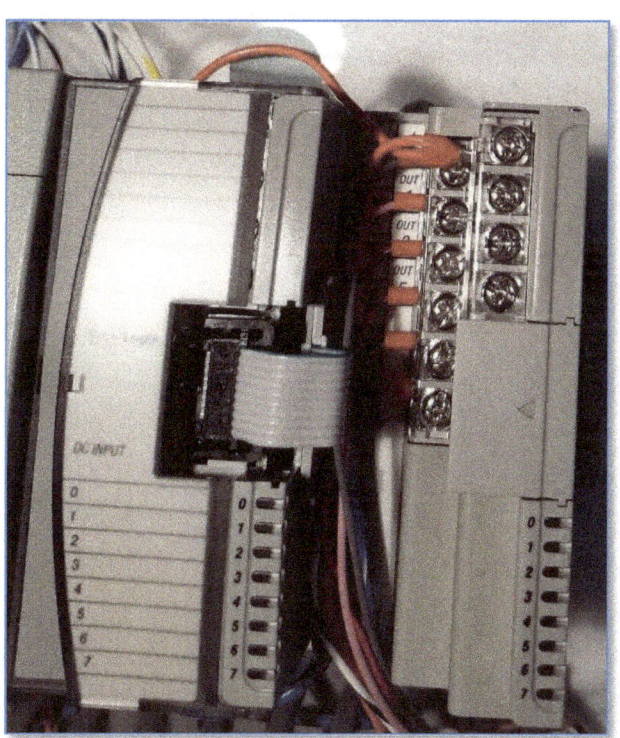

Slot 2 Bottom Row and completed terminations, Step 38.

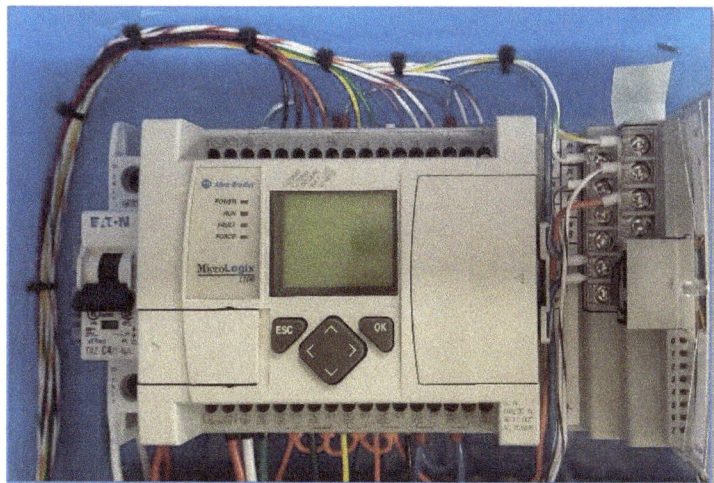

As sections of wiring on the PLC are completed, tie-wraps can be used to dress the wires as you go. Remember not to cut tie-wraps until they are properly tightened!

PLC chassis completed terminations, Step 38.

39. Final Wire Dressing - After completing all wiring terminations, use the tie-wraps (A033) to secure any loose wires and ensure that everything looks nice. Wrap the spiral wrap that you removed from the wiring bundle (A027) to control the wiring that passes from the door to the enclosure.

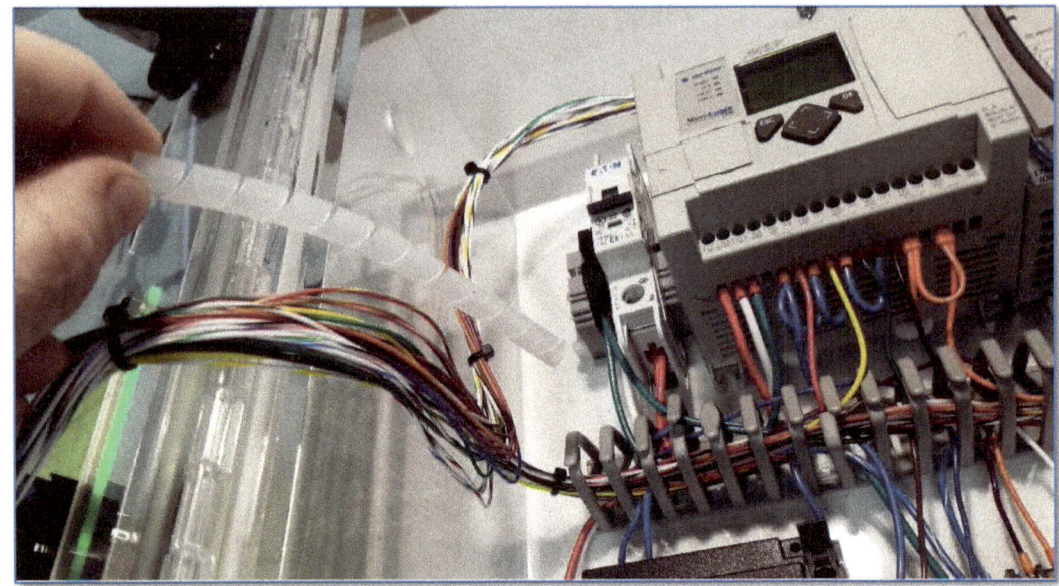

Spiral Wrap Measurement, Step 39.

Cut the spiral wrap to the correct desired length, from where the wire splits into two sections next to the wireway up to where the bundle splits on the door. Wrap the spiral wrap around the wires, this is where you can "hide" some of your length mismatches.

Spiral Wrap Completed, Step 39.

40. Power Cable – Insert the 3 prong pigtail (A012) into the smaller hole at the top left of the enclosure. Strip the outer insulation off of the cable so that about 3" of the wires inside are exposed. Strip the ends of the green, white and black wires and terminate them as sown in the figure below; the green is ground, the white is neutral, and the black is 120V power.

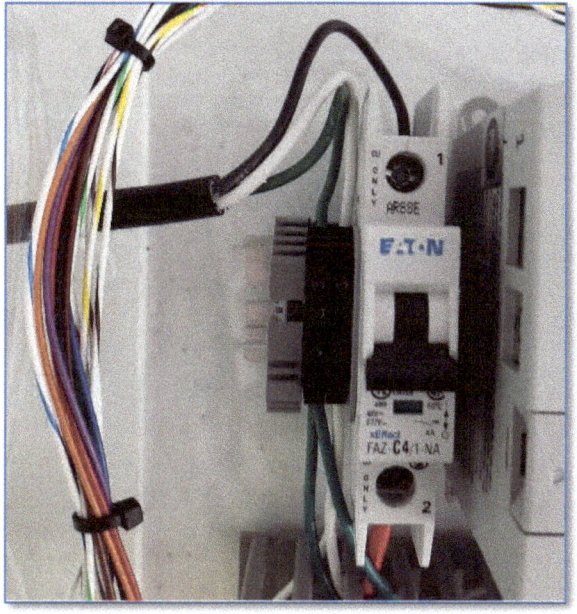

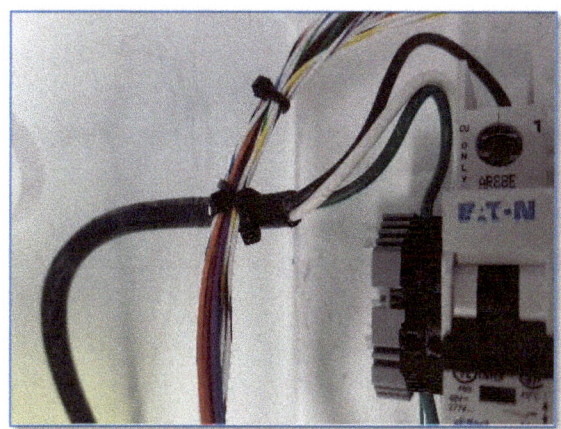

Pigtail (power) termination and strain relief, Step 40.

Wrap a tie-wrap (A033) around the cable just inside the enclosure to prevent the cable from being pulled out, this will act as a strain relief.

## Check and Apply Power

Before plugging your trainer in, there are several basic checks that can be done to ensure that there are no major problems with your trainer.

Carefully examine your wiring and ensure that all terminations are as explained in this document. Lightly pull on the wires to ensure that they are attached tightly to the terminals.

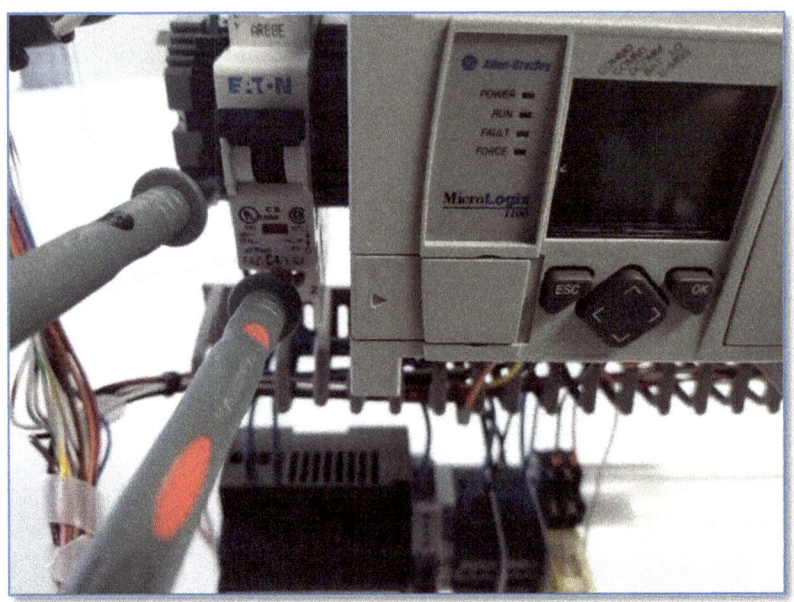

Using a multimeter, select "Ohms" for resistance measurement. With the circuit breaker in the on position, read the resistance between the bottom of the circuit breaker and the neutral terminal block. The resistance should be very high, when using a Micrologix 1100 (AC power) the reading was 0.5 MOhms. You are reading across the PLC and power supply's power terminals.

Close the fuse block and read from the +DC to the -DC terminals. The resistance should be about 600 Ohms (0.6kOhms). Using Ohms Law (I=V/R) this means the current is only about 40mA. Since the power supply is rated at 2.5A and the fuse is 3A, this is well within range.

Turn off the circuit breaker, and open the fuse block. Plug the power cord into an electrical outlet.

Switch the multimeter to the "Volts" setting, and AC if it is not autoranging, and read from the top of the circuit breaker to the neutral terminal. It should read 115-120 volts. Close the circuit breaker. The PLC lights should come on if it is AC powered. The power supply power light should also come on.

If there is a different selection for DC volts, select it on the meter. Read from the + to the – terminal on the power supply, it should read about 24 volts. Close the fuse block and read from the + terminal block to the – terminal blocks. It should still read 24 volts.

Your PLC should have indicators for the DC input terminals. Press the pushbuttons one at a time and ensure that the corresponding indicators come on. Do the same for the switches, they should be on in the up position.

Pull out the E-Stop pushbutton and press the Power Reset button. Ensure that the indicators for the MCR and E-Stop are on. When pushing the Power Reset button, its indicator should come on also. The power light should also be on with the MCR engaged.

To energize the lights, you will need to turn on the outputs on the PLC. This can only be done while online with the PLC. With no logic in the PLC (you still may need to load an empty program to go online), you can simply enter 1s in the output data table to turn on the outputs. Do this for each assigned output on the PLC and ensure that the corresponding indicator comes on and that the buzzer makes a sound. If the power light is on (MCR engaged) all lights should operate. If the E-Stop is pressed, all lights except for the stack lights should go off. The buzzer should still operate.

Look at the data register for the analog inputs. Turn the potentiometers to the far left and far right positions. If the registers are configured as integers, the left positions should give a reading of about 0 and the right about 32,000 or so.

At this point you have checked the functionality and wiring of your trainer. Now use it to practice your programming skills!

## About the Author

Frank Lamb is an industrial automation consultant and advanced PLC programming trainer with more than 30 years of experience in controls and machine automation.

From 1996 to 2006, Frank owned and operated Automation Consulting Services, Inc. (ACS), a panel building and machine integration company in Knoxville, TN. From 2006 to 2011, he worked as a senior-level project engineer for Wright Industries in Nashville, TN, where he led the design and implementation of large, complex systems and custom machines for multinational corporations and government agencies. In December 2011, Frank re-established Automation Consulting, LLC with a new vision: to use his experience in the field of industrial automation – from electrical, mechanical and controls engineering to project management, training, and machine documentation – to provide expert consulting and training services to manufacturers.

Frank is the president and owner of Automation Consulting, LLC in Nashville, TN and author of *Industrial Automation: Hands On*, published by McGraw-Hill Professional in 2013. He is a United States Air Force veteran, received his BSEE in Electrical and Computer Engineering from the University of Tennessee, and has a Green Belt in Lean Manufacturing/Six Sigma from Purdue University.